Martin Rudolph, M.S., M.A.
Former Chairman,
Mathematics Department
Oceanside High School
Oceanside, New York

Calculus

All inquiries should be addressed to:
Barron's Educational Series, Inc.
250 Wireless Boulevard
Hauppauge, New York 11788
http://www.barronseduc.com

Library of Congress Catalog Card No. 2001043940

International Standard Book No. 0-7641-1866-8

Library of Congress Cataloging-in-Publication Data
Rudolph, Martin, 1940–
    Calculus / Martin Rudolph.
      p.   cm. – (Barron's EZ-101 study keys)
    Includes index.
    ISBN 0-7641-1866-8
    1. Calculus. I. Title. II. Barron's EZ 101 study keys.
QA303.2 .R83 2002
515–dc21   2001043940

PRINTED IN THE UNITED STATES OF AMERICA
987654321

CONTENTS

Theme 1 PREREQUISITES

Welcome to the world of calculus. This book is intended to be used in conjunction with your calculus textbook. It is primarily devoted to giving clearly outlined, concise descriptions of *what* you need to know, along with many observations and examples explaining, step by step, *how* you can successfully execute the techniques of calculus. However, at critical points of the exposition, in order to ensure complete understanding, the question of *why* techniques work is briefly explored. In addition, this book contains many cross-references. When a key is noted, the reference is to the key itself or the notes/bulleted item(s) following the key. For the sake of brevity, student exercises are not offered; there are many calculus textbooks containing thousands of problems, some with solution manuals. Also, you can try the "key examples" in this book before reading their solutions.

As you know, mathematics is a cumulative subject, so you must be knowledgeable in precalculus math. In fact, the author has found that the major cause of difficulties in calculus is not calculus! Rather, it is a lack of sufficient knowledge of geometry, trigonometry, and especially algebra. Themes 1 and 2 serve as reminders of *some* of this math. If further details are needed about these or other topics, it is strongly recommended that you refer to a precalculus book such as Barron's *EZ-101 College Algebra.*

One more note: calculus textbooks do not agree about where to introduce trigonometric, logarithmic, and exponential functions. The author believes later is better because it permits a less complicated exposition of the basic concepts of calculus. Then, when the transcendental functions are considered, a necessary review of these concepts is automatically provided. Either way, calculus is one of humanity's most amazing inventions. Good luck, and enjoy the adventure.

Key 1 Algebra

OVERVIEW *Algebra is the language of mathematics. It is needed to read, write, and understand calculus. Keys 1-1 through 1-5 are examples of algebraic solutions. They are intended to indicate the depth and breadth of required knowledge as well as to provide a very brief review.*

Complex fractions: To simplify a complex fraction, multiply both the numerator and the denominator of the fraction by the least common denominator (LCD) of the fractions in the numerator and/or the denominator that make the fraction complex.

KEY EXAMPLE 1-1

Simplify the complex fraction: $\dfrac{\dfrac{1}{a^2} - \dfrac{2}{ab} + \dfrac{1}{b^2}}{\dfrac{1}{a^2} - \dfrac{1}{b^2}}$.

Solution:

$$\frac{\dfrac{1}{a^2} - \dfrac{2}{ab} + \dfrac{1}{b^2}}{\dfrac{1}{a^2} - \dfrac{1}{b^2}} = \frac{\dfrac{1}{a^2} - \dfrac{2}{ab} + \dfrac{1}{b^2}}{\dfrac{1}{a^2} - \dfrac{1}{b^2}} \cdot \frac{a^2 b^2}{a^2 b^2} \quad \text{(Multiply by LCD.)}$$

$$= \frac{b^2 - 2ab + a^2}{b^2 - a^2} \qquad \left(\begin{array}{l}\text{Don't forget the} \\ \text{distributive law.}\end{array}\right)$$

$$= \frac{(b-a)(b-a)}{(b+a)(b-a)} \qquad \left(\begin{array}{l}\text{Factor, and reduce} \\ \text{the fraction.}\end{array}\right)$$

$$= \frac{b-a}{b+a}$$

Solving polynomial equations: If the equation is first-degree, use standard techniques. If the equation is second-degree or higher, write the equation in standard form:

$$a_n x^n + a_{n-1} x^{n-1} + \ldots + a_2 x^2 + a_1 x + a_0 = 0$$

Factor the polynomial, and then set each factor equal to 0 and solve.

KEY EXAMPLE 1-2

Solve for x: $x^2 = 3x + 4$.

Solution:

$$x^2 = 3x + 4$$
$$x^2 - 3x - 4 = 0 \quad \text{(standard form)}$$
$$(x+1)(x-4) = 0 \quad \text{(factoring)}$$
$$x + 1 = 0 \quad \text{or} \quad x - 4 = 0 \quad (a \cdot b = 0 \implies a = 0 \text{ or } b = 0)$$
$$x = -1 \quad \text{or} \quad x = 4$$

- If the quadratic does not factor, use the quadratic formula:

$$x = \frac{-b \pm \sqrt{b^2 - 4ac}}{2a}$$

  to solve $ax^2 + bx + c = 0$, $a \neq 0$.

- When solving higher degree polynomial equations, factoring is the only simple method. You should review the remainder and factor theorems, the rational root theorem, and synthetic division.

Solving polynomial inequalities: If the degree of the polynomial equation is greater than 1, write the equation in standard form, factor the polynomial, and then draw a number line and analyze the situation. The following example assumes that the polynomial is already factored.

KEY EXAMPLE 1-3

Solve the inequality: $(x + 2)(x - 1)(x - 3)^2(x - 5) \geq 0$.

Solution: Reason as follows:

For all x in the interval	$x + 2$ is	$x - 1$ is	$(x - 3)^2$ is	$x - 5$ is	$(x + 2)(x - 1)(x - 3)^2(x - 5)$ is
$x < -2$	−	−	+	−	−
$-2 < x < 1$	+	−	+	−	+
$1 < x < 3$	+	+	+	−	−
$3 < x < 5$	+	+	+	−	−
$x > 5$	+	+	+	+	+

A number line helps to visualize the whole situation.

KEY DIAGRAM 1-4

(See table in KEY 1-3.)

NOTE: The polynomial is equal to 0 at $x = -2$, 1, 3, and 5.

$$\underset{-2}{\underline{-0}} \quad + \quad \underset{1}{\underline{0}} \quad - \quad \underset{3}{\underline{0}} \quad - \quad \underset{5}{\underline{0}} \quad + \quad (x+2)(x-1)(x-3)^2(x-5)$$

$$(x+2)(x-1)(x-3)^2(x-5) \geq 0 \quad \Rightarrow \quad -2 \leq x \leq 1 \quad \text{or} \quad x = 3 \quad \text{or} \quad x \geq 5$$

Solving inequalities: The rules are similar to those used to solve equations. When you add, subtract, multiply, or divide both sides of the inequality by the same number, the inequality maintains its "direction" *unless* you are multiplying or dividing by a *negative number*, in which case the inequality reverses its direction.

KEY EXAMPLE 1-5

Solve the inequality: $6 - 4x \leq 2x + 24$.

Solution:

$$6 - 4x \leq 2x + 24$$
$$6 - 6x \leq 24 \quad \text{(Subtract } 2x \text{ from both sides.)}$$
$$-6x \leq 18 \quad \text{(Subtract 6 from both sides.)}$$
$$x \geq -3 \quad \text{(Divide both sides by } (-6),$$
$$\text{and note change in direction.)}$$

Key 2   Sets and functions

OVERVIEW *A **set** is any well-defined collection of objects. You must be familiar with basic operations on sets such as intersection, union, and complementation. In calculus, we frequently refer to certain sets called **intervals** that have their own specialized notation. A **function** is a relationship between two sets, the **domain** and the **range**. Set and function are two fundamental concepts of mathematics. The following review is too brief to do them justice, so, if you have any doubts, read more about these concepts in any precalculus book.*

Set notation: We usually name sets with capital letters ($A,B,C, \ldots$), but certain important sets have special names. For example, $\mathbb{R}$ = the set of real numbers and $\emptyset$ = the empty (or null) set. The objects in a set are called *elements* or *members* of the set. There are two ways of indicating the members of a set. If a set is relatively small, we use the roster method: $S = \{-0.5, 3, 7, \sqrt{13}\}$. If a set is large or infinite, we use "set builder" notation: the set of all numbers between -1 and 4 inclusive is $A = \{x \in \mathbb{R} | -1 \leq x \leq 4\}$. The notation $3.5 \in A$ means that 3.5 is a member of set A, while $\sqrt{19} \notin A$ means that $\sqrt{19}$ is *not* a member of set A.

Convention: In mathematics, discussions are confined to a particular set, sometimes called the *universal set* or *domain*. If the universal set is not stated, then it is assumed to be $\mathbb{R}$. Therefore, we usually write set A, defined above, as $A = \{x | -1 \leq x \leq 4\}$. If set A contained only *integers*, we would write $A = \{x \in \mathbb{Z} | -1 \leq x \leq 4\}$, where $\mathbb{Z}$ = the set of integers.

Subset: Set A is a subset of set B ($A \subseteq B$) means that every element of A is an element of B. If $A \subseteq B$ and $B \subseteq A$, then $A = B$.

The intersection of two sets ($A \cap B$) is the set of all elements common to both.

KEY EXAMPLE 2-1

$A = \{1,2,3,4,5\}$ and $B = \{4,5,6\} \Rightarrow A \cap B = \{4,5\}$

The union of two sets ($A \cup B$) is the set of elements in either set or in both.

KEY EXAMPLE 2-2

$A = \{1,2,3,4\}$ and $B = \{4,5\} \Rightarrow A \cup B = \{1,2,3,4,5\}$

The complement of a set (A') is the set of elements in the universal set and *not* in the set itself.

KEY EXAMPLE 2-3

(NOTE: Since not stated explicitly, the universal set $= \mathbb{R}$.)

$A = \{x|x > 3\} \Rightarrow A' = \{x|x \leq 3\}$

Interval: For our purposes, an interval is any unbroken subset of a number line. The following table includes all possible intervals with two types of notation.

KEY TABLE 2-4

Set notation	Interval notation	Interval type	Graph
$\{x|a < x < b\}$	(a,b)	open	
$\{x|a \leq x \leq b\}$	$[a,b]$	closed	
$\{x|a < x \leq b\}$	$(a,b]$	half-open	
$\{x|a \leq x < b\}$	$[a,b)$	half-open	
$\{x|x > a\}$	(a,∞)	open ray	
$\{x|x \geq a\}$	$[a,\infty)$	closed ray	
$\{x|x < b\}$	$(-\infty,b)$	open ray	
$\{x|x \leq b\}$	$(-\infty,b]$	closed ray	
$\mathbb{R}$ or $\{x|x \in \mathbb{R}\}$	$(-\infty,\infty)$	entire line	

NOTE: Interval notation (a,b) assumes that $a < b$.

A **relation** matches the elements of two sets and is written as a set of ordered pairs. The set of all first coordinates is called the **domain**, and the set of all second coordinates is called the **range**.

NOTATION: A possible source of confusion is whether the symbol (a,b) represents an ordered pair of a relation or an open interval. The context should eliminate any ambiguities.

A **function** is a special relation in which each element in the domain is matched with *exactly one* element in the range. In other words, in a function, no two ordered pairs have the same first coordinate. The second coordinate of each ordered pair is called *the* **image** of the first coordinate, and the first coordinate is called *a* **pre-image** (there may be more than one) of the second. If every image has *exactly one* pre-image, then the function is 1-1 (one-to-one).

KEY DIAGRAM 2-5

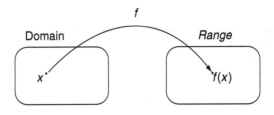

- If the name of a function is f, then $f(x)$ represents **the image of x**. It is also read as "f of x" or "f at x." The variable y is frequently used in place of $f(x)$. Since the value of y depends on the value of x, y is called the **dependent variable** and x is the **independent variable**.
- It may be helpful to imagine a function as sometimes acting like a slide projector. The domain is the slide itself, and the range is the *image* on the screen.
- To define a function completely, one needs to know the **rule(s) for calculating images** and the function's domain (the range can theoretically be determined from this information). **Convention:** If the domain of a function is not specified, it is assumed to be the set of all real numbers that are assigned real images by the rule. NOTE: The rule is not always simple; the *rule* may be an explicit set of ordered pairs.

KEY EXAMPLES 2-6

A. If $y = x^2$ and $x \geq 3$, then the domain is $\{x | x \geq 3\}$, as specified.

B. If $y = x^2$, then, by convention, the domain is $\mathbb{R}$ (set of all real numbers).

C. If $y = \sqrt{x-7}$, then, by convention, the domain is $\{x | x \geq 7\}$. Note that, if $x < 7$, then $\sqrt{x-7}$, the image of x, will be imaginary (not real).

- Since a function is a set of ordered pairs, the **graph** of a function is an indication (by "darkening") of those ordered pairs. Usually, the ordered pairs are represented by (x,y), but this is a generic representation for most functions. It is frequently beneficial to be more precise and to write $(x, f(x))$. In example C above, the ordered pairs of the given function can be represented by $(x, \sqrt{x-7})$. Then $\{(x, \sqrt{x-7}) | x \geq 7\}$ is a nice way to write the function.

A **piecewise function** is a function that has more than one rule for finding images, with each rule applicable to a different part of the domain. In the following example, $f(-2) = (-2)^2 = 4$, $f(1) = \dfrac{4}{1} = 4$, and $f(3) = \sqrt{3}$.

KEY EXAMPLE 2-7

What is the domain of the following piecewise function?

$$f(x) = \begin{cases} x^2 & \text{if } -3 \leq x \leq -1 \\ \dfrac{4}{x} & \text{if } -1 < x < 2 \\ \sqrt{x} & \text{if } x > 2 \end{cases}$$

Solution: The domain of f is $\{x | x \geq -3, x \neq 0, \text{ and } x \neq 2\}$. Make sure you understand the answer. Why are 0 and 2 excluded from the domain? (Try to find their images.)

KEY DEFINITION 2-8

An important piecewise function is the **absolute-value function**, which has the following definition:

$$|x| = \begin{cases} x & \text{if} \quad x \geq 0 \\ -x & \text{if} \quad x < 0 \end{cases} \qquad \text{The domain} = \mathbb{R}.$$

The absolute-value function plays an important role in the most basic calculus definitions. It is imperative that you have a complete knowledge of its algebraic properties. Be sure that you understand each of the following theorems.

KEY THEOREMS 2-9

A. $|x| \leq a, a > 0 \Rightarrow -a \leq x \leq a$
   $|x| < a, a > 0 \Rightarrow -a < x < a$
B. $|x| \geq a, a > 0 \Rightarrow x \geq a$ or $x \leq -a$
   $|x| > a, a > 0 \Rightarrow x > a$ or $x < -a$
C. $|x| = \sqrt{x^2}$
D. $|a - b| = |b - a|$
E. $|a||b| = |ab|$
F. $|a| + |b| \geq |a + b|$ (Triangle Inequality Theorem)

KEY EXAMPLE 2-10

Solve for x: $|2x + 4| > 2$.

Solution: Use KEY 2-9B; then
$$|2x + 4| > 2 \Rightarrow 2x + 4 > 2 \quad \text{or} \quad 2x + 4 < -2$$
$$2x > -2 \quad \text{or} \quad 2x < -6$$
$$x > -1 \quad \text{or} \quad x < -3$$

KEY EXAMPLE 2-11

Solve for x: $|x - a| < \delta, \delta > 0$.

Solution: Use KEY 2-9A; then

$$|x-a| < \delta, \delta > 0 \quad \Rightarrow \quad \begin{array}{c} -\delta < x - a < \delta \\ a - \delta < x < a + \delta \end{array}$$

KEY EXAMPLE 2-12

Solve for x: $|3x - 7| \geq -4$.

Solution: Every real number is a solution because $|3x - 7|$ is never negative, and, therefore, $|3x - 7| \geq -4$ for all x.

Algebra of functions: The definitions of addition, subtraction, multiplication, and division of functions are straightforward, so we provide only one example.

KEY EXAMPLE 2-13

If $f(x) = x^2$, and $g(x) = \sqrt{x-1}$, then

$$(f + g)(x) = f(x) + g(x) = x^2 + \sqrt{x-1}$$
$$(f - g)(x) = f(x) - g(x) = x^2 - \sqrt{x-1}$$
$$(f \cdot g)(x) = f(x)g(x) = x^2\sqrt{x-1}$$
$$(f \div g)(x) = \frac{f(x)}{g(x)} = \frac{x^2}{\sqrt{x-1}}$$

- The domain of the sum (or difference or product or quotient) of two functions is the *intersection* of their respective domains. This is true because, in order for x to be in any of the domains, both $f(x)$ and $g(x)$ have to exist; therefore, x must be in both domains. Of course, with division we would exclude any number that causes a denominator to be 0.

Composition of functions: Suppose a function f takes elements of set A (domain) and assigns them images in set B, and function g takes elements of B and assigns them images in set C. Then, using f and g, we can create a function that has A as its domain and C as its range. This function is called the *composition* (or *composite*) of f and g and is written as $g \circ f$. Its rule is $(g \circ f)(x) = g(f(x))$. The first function to operate, f, is sometimes called the *inner function*, and the last function, g, the *outer function*. The domain of a composite function must be determined *before* algebraic simplification. (See KEY 2-15.)

KEY DIAGRAM 2-14

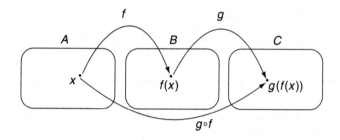

KEY EXAMPLE 2-15

If $f(x) = x^2 + 8$ and $g(x) = \sqrt{x}$, then

$$(g \circ f)(2) = g(f(2)) = g(2^2 + 8) = g(12) = \sqrt{12}$$
$$(f \circ g)(2) = f(g(2)) = f(\sqrt{2}) = (\sqrt{2})^2 + 8 = 10$$

In general:

$$(g \circ f)(x) = g(f(x)) = g(x^2 + 8) = \sqrt{x^2 + 8}$$
$$(f \circ g)(x) = f(g(x)) = f(\sqrt{x}) = (\sqrt{x})^2 + 8 = x + 8$$

- The domain of $f \circ g$ is $[0, \infty)$ even though the domain of $x + 8$ ($f \circ g$ after simplification) is $(-\infty, \infty)$. $(f \circ g)(-3)$ does *not* exist!
- The composition of functions is *not* <u>commutative</u> ($f \circ g \neq g \circ f$).
- Here is an important skill: Given $h(x) = \sqrt{x^2 + 8}$, to perceive $h(x)$

as the composite $g \circ f$ and to see clearly which function is the inner function and which is the outer function.

Inverse of a function: When the composite of two functions has the property that the image of every number in its domain is the same (original) number, then the two functions are called *inverse functions*.

KEY EXAMPLE 2-16

Consider the functions $f(x) = 5x$ and $g(x) = \frac{1}{5}x$.

$$(f \circ g)(x) = f(g(x)) = f\left(\frac{1}{5}x\right) = 5\left(\frac{1}{5}x\right) = x$$

$$(g \circ f)(x) = g(f(x)) = g(5x) = \frac{1}{5}(5x) = x$$

- In a sense, each of the two inverse functions seems to "cancel out" the effect of the other.
- Note that $f(2) = 10$ and $g(10) = 2$. Since functions are sets of ordered pairs, $(2,10) \in f$ and $(10,2) \in g$. From this example, it should be clear that the domain of a function equals the range of its inverse function and the range of a function equals the domain of its inverse.
- NOTATION: The inverse of f is usually denoted by f^{-1} and read as "inverse of f" or "f-inverse." Be careful not to confuse the "inverse notation" with "reciprocal notation." Although x^{-1} means $1/x$, f^{-1} does *not* mean $1/f$. To indicate the reciprocal of a function f, write $(f)^{-1}$.

KEY DIAGRAM 2-17

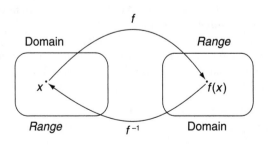

- Note again: $f^{-1}(f(x)) = x$ and $f(f^{-1}(x)) = x$.
- Recall that in the xy-plane points (a,b) and (b,a) are symmetric about the line $y = x$. If $(a,b) \in f$ means that $(b,a) \in f^{-1}$, then the graphs of $y = f(x)$ and $y = f^{-1}(x)$ are also symmetric about $y = x$.

Finding the inverse of a function: To find the inverse, you need to reverse the coordinates of all ordered pairs in the function.

- If f contains a finite number of ordered pairs, simply reverse the coordinates of each ordered pair. For example, if $f = \{(7,9), (0,1),(-3,2)\}$, then $f^{-1} = \{(9,7), (1,0), (2,-3)\}$.
- If f is defined by a rule, then, in the rule, replace x with y and y with x. Then, if possible, solve for y (getting an **explicit function**). Don't forget to interchange the domain and range.

KEY EXAMPLE 2-18

If

$$f = \{(x,y)|y = \sqrt{x-3}\} \quad \text{with} \quad \{x|x \geq 3\} \quad \text{and} \quad \{y|y \geq 0\}$$

then

$$f^{-1} = \{(x,y)|x = \sqrt{y-3}\} \quad \text{with} \quad \{x|x \geq 0\} \quad \text{and} \quad \{y|y \geq 3\}$$

Solve for y:

$$f^{-1} = \{(x,y)|y = x^2 + 3\} \quad \text{with} \quad \{x|x \geq 0\} \quad \text{and} \quad \{y|y \geq 3\}$$

- **WARNING**: Although $y = x^2 + 3$ is *the rule* for f^{-1}, it alone does not properly specify the inverse function itself because the domain of $y = x^2 + 3$ is (by convention) the set of all reals, but the domain of f^{-1} should be $\{x|x \geq 0\}$. **Always know the domains of the functions under discussion.**
- Switching the coordinates of all ordered pairs does not always result in a function (KEY 2-19). The inverse of a function is a function if and only if the original function is 1-1 (see text before KEY 2-5).

KEY EXAMPLE 2-19

If $f = \{(1,3), (2,3)\}$, then the inverse of f is $\{(3,1), (3,2)\}$, which is not a function. Recall that a function cannot assign two different images to the same number in its domain.

Often we require the inverse to be a function. In that case, we *restrict the domain* of the original function so that the inverse in the restricted domain *is* a function.

KEY EXAMPLE 2-20

Consider the function $y = x^2$ ($x \in \mathbb{R}$). Since $(9,3)$ and $(9,-3)$ are in the inverse, the inverse is not a function. Note that $y = x^2$ is not 1-1. However, if you restrict the domain of the function to $y = x^2$ ($x \geq 0$), then the inverse in that domain is a function (since $(-3,9) \notin f$, $(9,-3) \notin f^{-1}$).

Key 3 Geometry

OVERVIEW *Many calculus problems involve geometric relationships. Familiarity with formulas, such as those for area and volume, is vital. However, it is also important to know as many theorems of geometry as possible.*

Two-dimensional formulas: Symbols for measure are as follows: C = circumference; A = area; h = height or altitude; b, b_1, b_2 = base; and r = radius.

Rectangle and parallelogram: $A = bh$

Triangle: $A = \dfrac{1}{2} bh$

Circle: $A = \pi r^2$, $C = 2\pi r$

Trapezoid: $A = \dfrac{1}{2} h(b_1 + b_2)$

Three-dimensional formulas: Symbols for measure include the above and also V = volume; S = lateral surface area; B = area of base; and l = slant height.

Prism with parallel bases and rectangular box: $A = Bh$

Pyramid: $V = \dfrac{1}{3} Bh$

Right circular cone: $V = \dfrac{1}{3} \pi r^2 h$, $S = \pi r l$

Right circular cylinder: $V = \pi r^2 h$, $S = 2\pi r h$

Sphere: $V = \dfrac{4}{3} \pi r^3$, $S = 4\pi r^2$

Geometric relationships: The following is just a sampling of relevant theorems.

Theorem: If two lines intersect, then the vertical angles are congruent.
Theorem: If two sides of a triangle are congruent, then the angles opposite these sides are congruent, and conversely.
Theorem: If two parallel lines are cut by a transversal, then the alternate interior angles are congruent.

Theorem: If a line is tangent to a circle, then the radius drawn to the point of tangency is perpendicular to the tangent.

Theorem: In a circle, the measure of an inscribed angle is equal to one-half the measure of its intercepted arc.

Theorem of Pythagorus: In a right triangle, the square of the length of the hypotenuse equals the sum of the squares of the lengths of its legs.

Theorem: If two polygons are similar, then the corresponding sides (and other linear parts) are proportional (in the same ratio).

Theorem: If two angles of one triangle are congruent to two angles of a second triangle, then the triangles are similar.

Key 4 Trigonometry

OVERVIEW *It is essential to know the definitions of the trigonometric functions and their domains and ranges, as well as identities and formulas.*

If an angle is placed in a circle with its vertex at the center, then the **radian measure** of the angle is given by $\theta_{rad} = \dfrac{s}{r}$, where $s =$ the length of its intercepted arc. To convert between radian and degree measure, use $\dfrac{\theta_{rad}}{\pi} = \dfrac{\theta_{deg}}{180}$. The angle in a circle is said to be in **standard position** if its vertex is at the origin of a coordinate axis system and one of its sides is on the positive half of the *x*-axis, as shown in Key 4-1.

KEY DIAGRAM 4-1

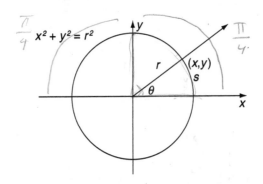

KEY DEFINITIONS 4-2

Use KEY 4-1; then

Definition	Domain (radians)	Range
$\sin\theta = \dfrac{y}{r}$	$-\infty < \theta < \infty$	$-1 \le \sin\theta \le 1$
$\cos\theta = \dfrac{x}{r}$	$-\infty < \theta < \infty$	$-1 \le \cos\theta \le 1$
$\tan\theta = \dfrac{y}{x}$	$\theta \ne (2k+1)\dfrac{\pi}{2}, k \in \mathbb{Z}$	$-\infty < \tan\theta < \infty$
$\cot\theta = \dfrac{x}{y}$	$\theta \ne k\pi, k \in \mathbb{Z}$	$-\infty < \cot\theta < \infty$
$\sec\theta = \dfrac{r}{x}$	$\theta \ne (2k+1)\dfrac{\pi}{2}, k \in \mathbb{Z}$	$\sec\theta \le -1$ or $\sec\theta \ge 1$
$\csc\theta = \dfrac{r}{y}$	$\theta \ne k\pi, k \in \mathbb{Z}$	$\csc\theta \le -1$ or $\csc\theta \ge 1$

To find the trigonometric function value of a number (angle) in its domain, we need to know (x,y), the coordinates of the point of intersection of the terminal side of the angle in standard position with a circle of known radius r. If the angle is in quadrant II, III, or IV, the problem can be reduced to finding the function value of its *reference angle* (the acute angle formed by the terminal side and the x-axis) and then assigning a plus or minus sign, depending on the trigonometric function and the quadrant. Although this is all done automatically (and very accurately) by calculators, there are certain *exact* values with which you are expected to be familiar. You might consider committing to memory the following table of exact trigonometric values for certain reference angles. Recall that cotangent, secant, and cosecant are reciprocals of the functions in the table.

	Reference angle		
Function	$\dfrac{\pi}{6}$	$\dfrac{\pi}{4}$	$\dfrac{\pi}{3}$
sin	$\dfrac{1}{2}$	$\dfrac{\sqrt{2}}{2}$	$\dfrac{\sqrt{3}}{2}$
cos	$\dfrac{\sqrt{3}}{2}$	$\dfrac{\sqrt{2}}{2}$	$\dfrac{1}{2}$
tan	$\dfrac{\sqrt{3}}{3}$	1	$\sqrt{3}$

KEY EXAMPLE 4-3

Find the value of $\sin \dfrac{5\pi}{3}$.

Solution: Draw a diagram.

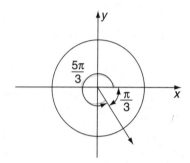

The diagram indicates that $\dfrac{5\pi}{3}$ is in quadrant IV and its reference angle is $\dfrac{\pi}{3}$. Because the sine function is *negative* in quadrant IV,

$$\sin \frac{5\pi}{3} = -\sin \frac{\pi}{3} = -\frac{\sqrt{3}}{2}.$$

KEY EXAMPLE 4-4

Find the value of $\csc \dfrac{5\pi}{3}$.

Solution: $\csc \dfrac{5\pi}{3} = \dfrac{1}{\sin \dfrac{5\pi}{3}} = -\dfrac{2}{\sqrt{3}} = -\dfrac{2\sqrt{3}}{3}$

An **identity** is an equation that is true for all permissible substitutions of the variables, that is, all numbers in the domain of the equation. The following list contains some of the more useful identities. Each identity has different algebraic forms, each of which may help with a problem. For example,

$$\sin^2 x + \cos^2 x = 1 \quad \Rightarrow \quad \sin^2 x = 1 - \cos^2 x$$

Fundamental Identities

$$\sin\theta\csc\theta = 1, \quad \cos\theta\sec\theta = 1, \quad \tan\theta\cot\theta = 1$$

$$\tan\theta = \frac{\sin\theta}{\cos\theta}, \quad \cot\theta = \frac{\cos\theta}{\sin\theta}$$

$$\sin^2\theta + \cos^2\theta = 1, \quad 1 + \tan^2\theta = \sec^2\theta, \quad 1 + \cot^2\theta = \csc^2\theta$$

Negative-Angle Identities

$$\sin(-\theta) = -\sin\theta, \qquad \cos(-\theta) = \cos\theta, \quad \tan(-\theta) = -\tan\theta,$$
$$\csc(-\theta) = -\csc\theta, \qquad \sec(-\theta) = \sec\theta, \quad \cot(-\theta) = -\cot\theta$$

Double-Angle Identities (with important variations)

$$\sin 2\theta = 2\sin\theta\cos\theta$$

$$\cos 2\theta = \cos^2\theta - \sin^2\theta$$

$$= 1 - 2\sin^2\theta \quad \Rightarrow \quad \sin^2\theta = \frac{1}{2}(1 - \cos 2\theta)$$

$$= 2\cos^2\theta - 1 \quad \Rightarrow \quad \cos^2\theta = \frac{1}{2}(1 + \cos 2\theta)$$

Sum and Difference Identities

$$\sin(a \pm b) = \sin a\cos b \pm \cos a\sin b$$

$$\cos(a \pm b) = \cos a\cos b \mp \sin a\sin b$$

$$\tan(a \pm b) = \frac{\tan a \pm \tan b}{1 \mp \tan a\tan b}$$

There are three pairs of cofunctions: sin and cos, tan and cot, sec and csc.

Law of Cofunctions: The function of any angle equals the cofunction of its complement. For example, $\cos\left(\dfrac{\pi}{2} - \theta\right) = \sin\theta$.

Law of Sines and Law of Cosines: In any $\triangle ABC$, a, b, and c represent the lengths of the sides of the triangle opposite angles A, B, and C, respectively.

$$\frac{a}{\sin A} = \frac{b}{\sin B} = \frac{c}{\sin C} \quad \text{and} \quad c^2 = a^2 + b^2 - 2ab\cos C$$

Even in this era of the graphing calculator, you should also be familiar with the basic graphs of the six trigonometric functions and their variations. For example, in $y = a \sin(bx + c) + d$, you should understand that a affects amplitude, b affects frequency, c affects horizontal displacement, and d affects vertical displacement.

KEY EXAMPLE 4-6

Solve $\cos 2x + 4 \sin x = -5$ for $0 \leq x \leq 2\pi$.

Solution:

$$\cos 2x + 4 \sin x = -5$$
$$1 - 2 \sin^2 x + 4 \sin x = -5$$
$$2 \sin^2 x - 4 \sin x - 6 = 0$$
$$\sin^2 x - 2 \sin x - 3 = 0$$
$$(\sin x + 1)(\sin x - 3) = 0$$

$\sin x + 1 = 0 \quad$ or $\quad \sin x - 3 = 0$

$\sin x = -1 \quad$ or $\quad \sin x = 3 \quad$ (Reject because $-1 \leq \sin x \leq 1$.)

$$x = \frac{3\pi}{2}$$

The solution set of the equation in the given domain is $\left\{ \dfrac{3\pi}{2} \right\}$.

Theme 2 ANALYTIC GEOMETRY

*E*very field of study has watersheds, at which knowledge takes a quantum leap forward. In mathematics, four of these moments occurred ca. 300 B.C., Euclid's *Elements* (axiomatic geometry); in 830, al-Khowarizmi's *Al-jabr w'al muqābalah* (algebra); in 1579, François Viète's *Canon mathematicus* (algebra and trigonometry); and in 1637, René Descartes's *La géométrie* (analytic geometry). The last of these provides mathematicians with a powerful tool, the wedding of geometry and algebra. Specifically, it gives us a method for visualizing algebraic equations by drawing **graphs** on a **coordinate plane**. This theme contains some of the basics of graphing and of the specific graphs of straight lines and the conic sections (circle, ellipse, hyperbola, and parabola).

Key 5   Coordinates; two important
formulas; locus

OVERVIEW *Here is a simple idea that revolutionized mathematics: Create a **coordinate (Cartesian) plane** by placing two number lines, called the **x-axis** and **y-axis**, perpendicular to each other at their origins. Assign to each point in the plane an **ordered pair of numbers**, called **coordinates** and representing the point's directed distances from the two axes. Now it is possible to represent a geometric figure with an algebraic equation (or inequality) and vice versa.*

Coordinates: Every point has a name, an ordered pair usually written as (x,y). The first coordinate, x, indicates the *horizontal* displacement of the point from the y-axis. The second coordinate, y, indicates the point's *vertical* displacement from the x-axis.

KEY DIAGRAM 5-1

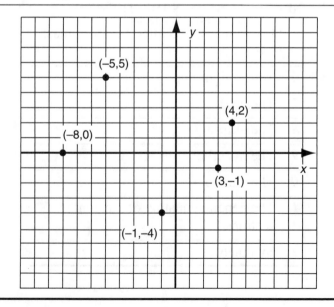

Two important formulas: Consider line segment $\overline{AB}$ with endpoints $A(x_1,y_1)$ and $B(x_2,y_2)$.

KEY DIAGRAM 5-2

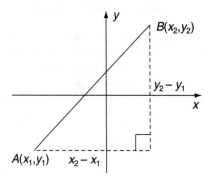

Distance formula: The distance between points A and B, that is, the length of $\overline{AB}$, is given by

$$AB = \sqrt{(x_2 - x_1)^2 + (y_2 - y_1)^2}$$

Midpoint formula: The midpoint of $\overline{AB}$ is $\left(\dfrac{x_1 + x_2}{2}, \dfrac{y_1 + y_2}{2} \right)$.

KEY EXAMPLE 5-3

Given $A(-3,-1)$ and $B(-7,5)$, find (a) AB and (b) the midpoint of $\overline{AB}$.

Solutions: (a) Use the distance formula; then

$$AB = \sqrt{[(-3) - (-7)]^2 + [(-1) - (-5)]^2} = \sqrt{(4)^2 + (-6)^2} = \sqrt{52} = 2\sqrt{13}$$

(b) Use the midpoint formula; then the midpoint of $\overline{AB}$ is

$$\left(\frac{(-3) + (-7)}{2}, \frac{(-1) + (5)}{2} \right) = (-5, 2)$$

KEY DEFINITION 5-4

A **locus of points** is a set of points that satisfy certain given conditions. A description of a locus of points means that every point in the set satisfies the given conditions *and* every point not in the set does not satisfy the given conditions.

A graph as a locus: The graph of an equation (e.g., $y = x^2$) is a picture of the locus of points (indicated by darkening the points) whose coordinates (x,y) satisfy the equation (in this case, the second coordinate equals the square of the first coordinate). In other words, every point on the graph satisfies the equation *and* every point not on the graph does not.

Key 6 The straight line

OVERVIEW *In the Cartesian plane, an important characteristic of a geometric figure is its "steepness" as* **x** *increases, which is indicated by a number called the* **slope**. *A (straight) line has the property that its slope remains constant; that is, its steepness does not change.*

KEY DEFINITION 6-1

The slope, m, of a nonvertical line passing through points $P(x_1,y_1)$ and $Q(x_2,y_2)$ is given by

$$m = \frac{\text{(vertical) change in } y}{\text{(horizontal) change in } x} = \frac{\Delta y}{\Delta x} = \frac{y_2 - y_1}{x_2 - x_1}$$

- Since the slope of a line is constant, *any* two points on the line can be used to calculate the slope.
- If a line is horizontal (parallel to the x-axis), then $\Delta y = 0$. Therefore, the slope of a horizontal line is 0.
- If a line is vertical (parallel to the y-axis), then $\Delta x = 0$. In that case, the line *has no slope* (or the slope is undefined).
- It should be clear that (1) if $m > 0$, the graph of the line goes from lower left to upper right, and (2) if $m < 0$, the graph goes from upper left to lower right. Also, the larger the magnitude of the slope, the "steeper" the graph.
- The points of intersection of the graph with the x-axis and y-axis are called the **x-intercept** and **y-intercept**, respectively.
- Mathematicians use the Greek capital letter delta (Δ) to indicate a change (frequently small) in the value of a variable ($\Delta x = $ a change in x).

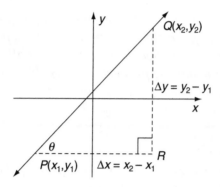

- In right triangle PRQ, $\tan\theta = \dfrac{\Delta y}{\Delta x}$. Therefore, $m = \tan\theta$. Angle θ is called the **angle of inclination** of the line.
- Parallel lines have equal slopes (or both slopes are undefined).
- Nonvertical perpendicular lines have slopes that are *negative reciprocals*; that is, the product of their slopes equals negative 1. For example, $(-3)\left(+\dfrac{1}{3}\right) = -1.$
- In many books, Δy is called "the rise" and Δx "the run."
- The calculations involving midpoints, distances, and slopes have many similarities; be careful not to confuse the formulas.

KEY EXAMPLE 6-3

Find the slope of the line passing through points $P(1,-6)$ and $Q(-3, 4)$.

Solution: $m = \dfrac{\Delta y}{\Delta x} = \dfrac{y_2 - y_1}{x_2 - x_1} = \dfrac{(4)-(-6)}{(-3)-(1)} = \dfrac{10}{-4} = -\dfrac{5}{2}$

NOTE: The slope of any line perpendicular to $\overleftrightarrow{PQ}$ will equal $\left(+\dfrac{2}{5}\right)$.

KEY FORMULAS 6-4

The equation of a line has many useful forms, each of which has a name as well as advantages and disadvantages.

Formula name	Formula	Comments
point-slope	$y - y_1 = m(x - x_1)$	Use this form if you know the slope, m, and the coordinates of *any* fixed point (x_1, y_1) on the line.
slope-intercept	$y = mx + b$	Solved for y, this expresses y as an explicit function of x; b = y-intercept.
two-intercept	$\dfrac{x}{a} + \dfrac{y}{b} = 1$	a = x-intercept; b = y-intercept; use on those (rare) occasions when both intercepts are known.
vertical	$x = a$	a = x-intercept; there is no y-intercept.
horizontal	$y = b$	b = y-intercept; there is no x-intercept.
general or standard	$Ax + By = C$	A line equation is often written this way, but it is usually advisable to change to the slope-intercept form.

- To find the slope of a line: If two points on the line are known, use the slope formula (KEY 6-1). If the equation of the line is known, solve for y (slope-intercept form) and the coefficient of x will be the slope.
- To find the intercepts of a line: In the equation of the line, substitute 0 for x and solve for y(-intercept); substitute 0 for y and solve for x(-intercept).

KEY NOTE 6-5

In regard to the above formulas (and in similar circumstances), students are frequently confused about the *meanings* of the different variables. For example, in $y = mx + b$, x and y represent the coordinates of every point on the line and are (**true**) **variables**; but m and b, although representing any real numbers, are "fixed" in the discussion. For purposes of clarification, the author will refer to the latter as **"constant variables."** (This is *not* standard terminology.)

KEY EXAMPLE 6-6

Find the equation of the line passing through points $P(4,-1)$ and $Q(0,5)$.

Solution: First find the slope:

$$M_{\overline{PQ}} = \frac{\Delta y}{\Delta x} = \frac{(-1)-(5)}{(4)-(0)} = -\frac{3}{2}$$

Refer to the point-slope formula (use either P or Q—the result is the same!):

$$y+1 = -\frac{3}{2}(x-4) \implies y = -\frac{3}{2}x+5$$

$$y-5 = -\frac{3}{2}(x-0) \implies y = -\frac{3}{2}x+5$$

- In this problem, one of the two points, $(0,5)$, being on the y-axis, is the y-intercept of the line. Therefore, after finding the slope, we could have saved some time by using the slope-intercept formula and could have written the equation without any required calculations.

KEY EXAMPLE 6-7

Find the equation of the locus of points equidistant from points $A(-3,4)$ and $B(1,2)$.

Solution (geometric): From elementary geometry, the locus of points equidistant from two distinct points is the perpendicular bisector of their segment. To find the equation of the perpendicular bisector, you need the slope and a point on the line. The slope of a perpendicular line will be the negative reciprocal of the slope of $\overline{AB}$, and the midpoint of the segment will be on the bisector.

$$m_{\overline{AB}} = \frac{\Delta y}{\Delta x} = \frac{(4)-(2)}{(-3)-(1)} = -\frac{1}{2} \implies m = 2 \quad \text{(KEY 6-1)}$$

$$\left[\text{Midpoint of } \overline{AB}\right] = \left(\frac{(-3)+(1)}{2}, \frac{(4)+(2)}{2}\right) = (-1,3)$$

By the point-slope formula, the equation of the locus is

$$y-3 = 2(x+1) \quad \text{or} \quad y = 2x+5$$

Solution (locus): Letting $P(x,y)$ represent the coordinates of any point on the locus, you need to find an equation (relating x and y) such that point P is equidistant from points A and B (i.e., $PA = PB$). Use the distance formula; then

$$PA = PB$$
$$\sqrt{(x+3)^2 + (y-4)^2} = \sqrt{(x-1)^2 + (y-2)^2}$$
$$(x+3)^2 + (y-4)^2 = (x-1)^2 + (y-2)^2$$
$$\cancel{x^2} + 6x + 9 + \cancel{y^2} - 8y + 16 = \cancel{x^2} - 2x + 1 + \cancel{y^2} - 4y + 4$$
$$-4y = -8x - 20$$
$$y = 2x + 5$$

Key 7 The circle

OVERVIEW *Every point on a circle is the same distance from the center of the circle as every other point.*

KEY DEFINITIONS 7-1

A **circle** is the locus of points in a plane that are a given distance from a given point. The given distance is called the **radius**, and the given point is called the **center**. The circle is said to be in **standard position** if and only if its center is at the origin, (0,0).

KEY DIAGRAM 7-2

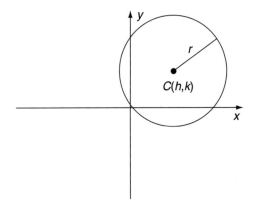

KEY FORMULA 7-3

The **standard form** of the equation of a circle with radius $= r$ and center at (h,k) is $(x - h)^2 + (y - k)^2 = r^2$.

KEY EXAMPLE 7-4

Write the equation of the locus of points 5 units from point (4,−3).

Solution: Use KEY 7-3; then

$$(x-4)^2 + (y+3)^2 = 25$$

KEY EXAMPLE 7-5

Write the equation of the circle that has a diameter with endpoints at $A(1,2)$ and $B(-3,4)$.

Solution: Use the midpoint formula to find the center, M, and the distance formula to find MA, the length of the radius.

$$\text{Midpoint of } \overline{AB} = \left(\frac{(1)+(-3)}{2}, \frac{(2)+(4)}{2} \right) = (-1,3), \text{ the center}$$

$$MA = \sqrt{(-1-1)^2 + (3-2)^2} = \sqrt{5}, \text{ the radius}$$

Therefore, the equation of the circle is $(x + 1)^2 + (y - 3)^2 = 5$.

KEY EXAMPLE 7-6

Write the equation of the circle whose center is at $C(3,4)$ and that is tangent to the x-axis.

Solution: Since the center is given, you need only the length of the radius. From elementary geometry, you know that, if a line (the x-axis) is tangent to a circle, then the radius drawn to the point of tangency is perpendicular to the tangent (the x-axis)(KEY 3). Since that radius is perpendicular to the x-axis, it measures the distance of the center from the x-axis. That distance is given by the y-coordinate of the center, so the radius of the circle is 4, as shown in the accompanying diagram.

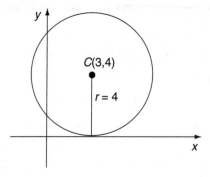

Therefore, the equation of the circle is $(x - 3)^2 + (y - 4)^2 = 16$.

KEY EXAMPLE 7-7

If the equation of a circle is $x^2 + y^2 - 4x + 6y - 23 = 0$, find its center and radius.

Solution: By *completing the square*, put the equation into **standard form**:

$$x^2 + y^2 - 4x + 6y - 23 = 0$$
$$(x^2 - 4x) + (y^2 + 6y) = 23$$
$$(x^2 - 4x + \boxed{4}) + (y^2 + 6y + \boxed{9}) = 23 + \boxed{4} + \boxed{9}$$
$$(x - 2)^2 + (y + 3)^2 = 36$$

Now, you can see that the graph is a circle whose center is at $(2, -3)$ and that $r = 6$.

The degenerate case: If, on completing the square, the "right-hand side" of the equation equals 0, for example, $(x - 5)^2 + (y + 6)^2 = 0$, then the graph is a single point, $(5, -6)$, the center. If the right-hand side is negative, for example, $(x - 5)^2 + (y + 6)^2 = -4$, then the solution set degenerates to the empty set.

Key 8 The ellipse

OVERVIEW *An ellipse is determined by two fixed points and a positive number representing the sum of the distances to these points.*

KEY DEFINITIONS 8-1

An **ellipse** is the locus of points in a plane the sum of whose distances to two fixed points is constant. The two fixed points, F_1 and F_2, are called **foci** (plural of *focus*), and the constant is represented by $2a$ (rather than a, $2a$ is chosen so that the equation of the ellipse is simplified). The midpoint of $\overline{F_1F_2}$ is the **center**, (h,k), of the ellipse. The ellipse is in **standard position** if and only if its center is at the origin and its foci are on one of the coordinate axes. The line segment passing through the foci with endpoints on the ellipse is the **major axis**. The line segment perpendicular to the major axis at the center with endpoints on the ellipse is the **minor axis**. The endpoints of the major axis are **vertices** (plural of *vertex*). The lengths of the major and minor axes are $2a$ and $2b$, respectively. Therefore, the distance of each vertex from the center is a. We represent the distance of each focus from the center by c. Disregarding rotations, we can say that there are two types of ellipses—those with horizontal major axes and those with vertical major axes.

KEY DIAGRAM 8-2

An ellipse with a horizontal major axis

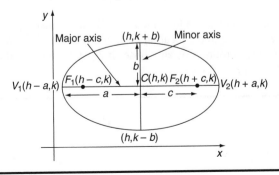

KEY DIAGRAM 8-3

An ellipse with a vertical major axis

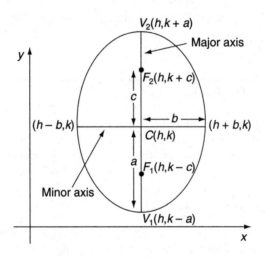

KEY FORMULAS 8-4

In both of the following equations, $b^2 = a^2 - c^2$, $a > b > 0$, and $a > c > 0$.

A. Ellipse with horizontal major axis: $\dfrac{(x-h)^2}{a^2} + \dfrac{(y-k)^2}{b^2} = 1$

B. Ellipse with vertical major axis: $\dfrac{(x-h)^2}{b^2} + \dfrac{(y-k)^2}{a^2} = 1$

- The two equations look identical. To distinguish between them, be sensitive to the location of a^2 and recall that $a^2 > b^2$.
- By the definition of an ellipse, if $P(x,y)$ is on the ellipse, then $PF_1 + PF_2 = 2a$.

KEY EXAMPLE 8-5

If the equation of an ellipse is $\dfrac{x^2}{4}+\dfrac{(y-4)^2}{9}=1$, find (a) the coordinates of the center, vertices, and foci, and (b) the lengths of the major and minor axes.

Solutions: (a) By observation, $a^2=9$ and $b^2=4$. Because of the position of a^2, the ellipse has a vertical major axis.

$$b^2=a^2-c^2 \quad \Rightarrow \quad c^2=5 \quad \Rightarrow \quad c=\sqrt{5}$$

Therefore, the ellipse has center $(0,4)$; $a=3 \Rightarrow$ vertices $(0,7)$ and $(0,1)$; and $c=\sqrt{5} \Rightarrow$ foci $(0,4+\sqrt{5})$ and $(0,4-\sqrt{5})$.

(b) The length of the major axis is 6, and the length of the minor axis is 4.

KEY EXAMPLE 8-6

Write the equation of the ellipse with vertices $(2,-1)$ and $(8,-1)$, and foci $(3,-1)$ and $(7,-1)$.

Solution: Since the foci (and vertices) lie on a horizontal line, the ellipse has a horizontal major axis. Therefore, its equation has the form

$$\frac{(x-h)^2}{a^2}+\frac{(y-k)^2}{b^2}=1$$

The center is the midpoint of the segment joining the foci: $(5,-1)$. Therefore, $h=5$ and $k=-1$. The distance of each vertex from the center $=a=3$. The distance of each focus from the center $=c=2$. Since $b^2=a^2-c^2$, $b^2=5$. Using this information, write the equation of the ellipse:

$$\frac{(x-5)^2}{9}+\frac{(y+1)^2}{4}=1$$

The degenerate case: If, on completing the square, the "right-hand side" of the equation equals 0, for example, $\dfrac{(x-2)^2}{25}+\dfrac{(y+9)^2}{13}=0$, then the graph is a single point, $(2,-9)$, the center. If the right-hand side is negative, then the solution set degenerates to the empty set.

Key 9   The hyperbola

OVERVIEW *A hyperbola is determined by two fixed points and a positive number representing the difference of the distances to these points.*

KEY DEFINITIONS 9-1

A **hyperbola** is the locus of points in a plane the absolute value of the difference of whose distances to two fixed points is constant. The two fixed points, F_1 and F_2, are called **foci**, and the constant is represented by $2a$ (rather than a, $2a$ is chosen so that the equation of the hyperbola is simplified). The midpoint of $\overline{F_1F_2}$ is the **center**, (h,k), of the hyperbola. The hyperbola is in **standard position** if and only if its center is at the origin and its foci are on one of the coordinate axes. The line segment with endpoints at the points of intersection of the hyperbola with the line containing the foci is the **transverse axis**. The endpoints of the transverse axis are **vertices**. The length of the transverse axis is $2a$. Therefore, the distance of each vertex from the center is a. We represent the distance of each focus from the center by c. Disregarding rotations, we can say that there are two types of hyperbolas, those with horizontal transverse axes and those with vertical transverse axes.

KEY DIAGRAM 9-2

A hyperbola with a horizontal transverse axis

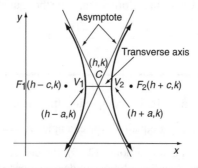

A hyperbola with a vertical transverse axis

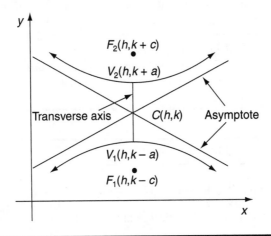

In both of the following equations, $b^2 = c^2 - a^2$, $c > b > 0$, and $c > a > 0$.

A. Hyperbola with horizontal major axis: $\dfrac{(x-h)^2}{a^2} - \dfrac{(y-k)^2}{b^2} = 1$

B. Hyperbola with vertical major axis: $\dfrac{(y-k)^2}{a^2} - \dfrac{(x-h)^2}{b^2} = 1$

Every hyperbola has two **asymptotes**. The equations of the asymptotes for the two types of hyperbolas defined above are:

A. For a hyperbola with horizontal major axis: $y - k = \pm\dfrac{b}{a}(x-h)$

B. For a hyperbola with vertical major axis: $y - k = \pm\dfrac{a}{b}(x-h)$

KEY FORMULA 9-6

There is one important hyperbola that is different from the above two types because it has its center at the origin but is *rotated* 45° out of standard position. The equation of this hyperbola is $xy = k$. If $k > 0$, then the foci are on the line $y = x$ and the graph of the hyperbola lies entirely in the first and third quadrants. If $k < 0$, then the foci are on the line $y = -x$ and the graph lies entirely in the second and fourth quadrants. In both cases, the asymptotes are $x = 0$ and $y = 0$ (the y-axis and x-axis).

KEY EXAMPLE 9-7

If the equation of a hyperbola is $\dfrac{(x-2)^2}{9} - \dfrac{(y-5)^2}{16} = 1$, find (a) the coordinates of the center, vertices, and foci, and (b) the equations of its asymptotes.

Solutions: (a) By observation, $a^2 = 9$ and $b^2 = 16$. Because of the position of a^2, the hyperbola has a horizontal transverse axis.

$$b^2 = c^2 - a^2 \implies c^2 = 25 \implies c = 5$$

Therefore, the hyperbola has center (2,5); $a = 3 \implies$ vertices (2,8) and (2,2); and $c = 5 \implies$ foci (2,10) and (2,0).

(b) The equations of the asymptotes are $y - 5 = \pm\dfrac{4}{3}(x-2)$.

KEY EXAMPLE 9-8

Write the equation of the hyperbola with vertices (2,–1) and (2,–7), and foci (2,1) and (2,–9).

Solution: Since the foci (and vertices) lie on a vertical line, the hyperbola has a vertical transverse axis. Therefore, its equation has the form

$$\frac{(y-k)^2}{a^2} - \frac{(x-h)^2}{b^2} = 1$$

The center is the midpoint of the segment joining the foci: (2,–4). Therefore, $h = 2$ and $k = -4$. The distance of each vertex from the center $= a = 3$. The distance of each focus from the center $= c = 5$. Since $b^2 = c^2 - a^2$, $b^2 = 16$. Using this information, write the equation of the hyperbola:

$$\frac{(y+4)^2}{9} - \frac{(x-2)^2}{16} = 1$$

The degenerate case: If, on completing the square, the "right-hand side" of the equation equals 0, for example, $\dfrac{(x-2)^2}{25} - \dfrac{(y+9)^2}{49} = 0$, then the graph is a pair of lines passing through "the center" [(2,−9)]. Their equations are $y + 9 = \pm\dfrac{7}{5}(x-2)$. If the right-hand side is negative, then the equation does not degenerate. When both sides of the equation are multiplied by (−1), the graph is clearly a hyperbola.

Key 10 The parabola

OVERVIEW *Every point on a parabola is the same distance from a line and a point.*

KEY DEFINITIONS 10-1

A **parabola** is the locus of points in a plane that are *equidistant* from a line and a point not on the line. The given line is the **directrix**, and the given point is the **focus**. The line perpendicular to the directrix and passing through the focus is the **axis of symmetry**. The midpoint of the perpendicular segment from the focus to the directrix is the **vertex**, (h,k). The *directed distance* from vertex to focus is represented by p. Whether p is positive or negative determines the orientation of the parabola. The parabola is in **standard position** if its vertex is at the origin and its focus is on one of the coordinate axes. Disregarding rotations, we can say that there are two types of parabolas, those with vertical axes of symmetry and those with horizontal axes of symmetry.

KEY DIAGRAM 10-2

A pair of parabolas with vertical axes of symmetry

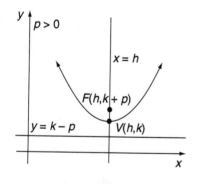
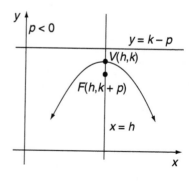

KEY DIAGRAM 10-3

A pair of parabolas with horizontal axes of symmetry

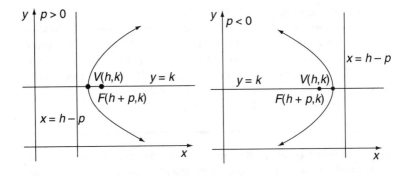

KEY FORMULAS 10-4

In both of the following equations, p represents the *directed distance* **from** vertex **to** focus.

A. Parabola with vertical axis of symmetry: $(x - h)^2 = 4p(y - k)$
B. Parabola with horizontal axis of symmetry: $(y - k)^2 = 4p(x - h)$

KEY EXAMPLE 10-5

If the equation of a parabola is $(y + 2)^2 = 24(x - 3)$, find (a) the coordinates of its vertex and focus, and (b) the equations of its directrix and axis of symmetry.

Solutions: (a) The form of the equation (KEY 10-4) indicates that the parabola has a *horizontal* axis of symmetry and, therefore, a vertical directrix. The vertex is at $(3,-2)$. Since $4p = 24$, then $p = 6$, the directed distance from vertex to focus. Therefore, the focus is +6 units (horizontally, to the right) from the vertex, at $(9,-2)$.

(b) The directrix is a vertical line 6 units to the left of the vertex with equation $x = -3$. The axis of symmetry is the horizontal line passing through the vertex and focus, $y = -2$.

KEY EXAMPLE 10-6

What is the equation of the locus of points equidistant from the line $y = 7$ and point $(1,3)$?

Solution: By definition (KEY 10-1), the graph is a parabola. Since its directrix is horizontal ($y = 7$), its axis of symmetry is vertical. By KEY 10-4, the equation will have the form $(x - h)^2 = 4p(y - k)$. The vertex, (h,k), is the midpoint of the perpendicular segment from the focus, $(1,3)$, to the directrix, at $(1,5)$. The directed (vertical) distance from vertex to focus is $p = -2$. Therefore, the equation of the locus is
$$(x - 1)^2 = -8(y - 5)$$

The degenerate case: In attempting to write the equation of a parabola in standard form, if one side of the equation equals 0, for example, $(y + 3)^2 = 0$, then the parabola degenerates to one line ($y = -3$).

Key 11   The general equation of the conic sections

OVERVIEW *It is possible to express the equations of all the conic sections and their degenerate cases by a single equation.*

KEY FORMULA 11-1

The general form of the equation of all conic sections is

$$Ax^2 + Bxy + Cy^2 + Dx + Ey + F = 0$$

- The values of A, B, C, D, E, and F determine the type of conic section, its location in the Cartesian plane, and its size/shape.
- $B^2 - 4AC$ is called the **discriminant**.
- If the conic section does not degenerate, then A, B, and C determine the type:
    If $B^2 - 4AC > 0$, the conic is a circle or an ellipse.
    If $B^2 - 4AC = 0$, the conic is a parabola.
    If $B^2 - 4AC < 0$, the conic is a hyperbola.
- If $B \neq 0$, then the conic is *rotated* out of standard position. This case is usually omitted from first-year calculus courses, except for a special case of the hyperbola mentioned in KEY 9-6.
- If $D \neq 0$, then the conic is *horizontally translated* out of standard position. If $E \neq 0$, then the conic is *vertically translated* out of standard position.

Three degenerate cases:

- If $B = D = E = F = 0$, $A > 0$, and $C > 0$, then the conic degenerates to a single point.
- If $B = D = E = F = 0$, $A > 0$, and $C < 0$, then the conic degenerates to a pair of straight lines.
- If $A = B = C = 0$ and $E \neq 0$ or $F \neq 0$, then the conic degenerates to one straight line.

Key 12 Symmetries

OVERVIEW *The graphs of conic sections exhibit two types of symmetries: about a line and about a point. Since symmetry (and asymmetry) is an important characteristic of a graph, we explore methods for recognizing each property.*

KEY DEFINITION 12-1

Two points, *A* and *B*, are **symmetric** to each other ("symmetric partners") about a *line* if and only if the line is the perpendicular bisector of their segment, $\overline{AB}$.

KEY DEFINITION 12-2

Two points, *A* and *B*, are **symmetric** to each other ("symmetric partners") about a *point P* if and only if *P* is the midpoint of their segment, $\overline{AB}$.

Special symmetries: While all symmetries are interesting, there are four special symmetries in analytic geometry with which we must be familiar: symmetries about the *x*-axis, the *y*-axis, the origin (0,0), and the line $y = x$.

KEY DIAGRAM 12-3

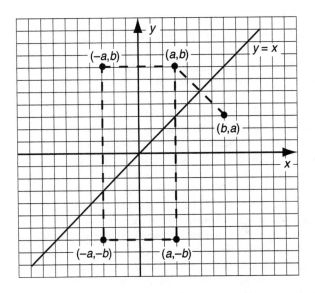

KEY OBSERVATIONS 12-4

The "symmetric partner" of point (a,b)

* about the x-axis is $(a,-b)$,
* about the y-axis is $(-a,b)$,
* about the origin is $(-a,-b)$,
* about $y = x$ is (b,a). (see KEY 2-17).

KEY DEFINITIONS 12-5

* A graph has a certain symmetry if and only if *every* point on the graph has its symmetric partner also on the graph.
* One graph has a certain symmetry to a second graph if and only if *every* point on each graph has its symmetric partner on the other graph.

KEY OBSERVATIONS 12-6

Concerning symmetry of a single graph

- **About the y-axis:** Point (a,b) on the graph implies that $(-a,b)$ is on the graph. Therefore, if x occurs to only even powers, then the graph must have y-axis symmetry. (NOTE: a raised to an even power equals $(-a)$ to the same even power.) Be careful; the converse is not true: $y = |x|$ *is* symmetric about the y-axis, but x occurs to an odd power. In addition, if the domain is asymmetric about $x = 0$ (the y-axis), then the graph must be asymmetric about the y-axis. NOTE: A function that is symmetric about the y-axis has the property $f(-x) = f(x)$ and is called an **even function**. Examples of even functions are $y = x^2$, $y = 7x^6$, $y = |x|$, and $y = \cos x$.

- **About the x-axis:** Point (a,b) on the graph implies that $(a,-b)$ is on the graph. Therefore, if y occurs to only even powers, then the graph must have x-axis symmetry. (NOTE: b raised to an even power equals $(-b)$ to the same even power.) Be careful; the converse is not true: $x = |y|$ *is* symmetric about the y-axis, but y occurs to an odd power. In addition, if the range is asymmetric about $y = 0$ (the x-axis), then the graph must be asymmetric about the x-axis.

- **About the origin:** Point (a,b) on the graph implies that $(-a,-b)$ is on the graph. Therefore, if both x and y occur to only even powers, then the graph must have symmetry about $(0,0)$. In addition, if either the domain or the range is asymmetric about 0, then the graph is asymmetric about the origin. NOTE: A function that is symmetric about the origin has the property $f(-x) = -f(x)$ and is called an **odd function**. Examples of odd functions are $y = x^3$, $y = 8x^5$, $y = \sqrt[3]{x}$, $y = \sin x$, and $y = \tan x$.

- If a graph has any two of these three symmetries, then it *must* have the third.

- An asymmetry can be verified by finding any single point on the graph whose symmetric partner is not on the graph.

KEY EXAMPLE 12-7

Discuss the symmetries of $y^2 = \dfrac{x+3}{x-2}$.

Solution: Consider each symmetry separately.

x-axis: Yes, since y occurs to only even powers. In addition, note that the range, $\{y \,|\, y \neq \pm 1\}$, is also symmetric about 0. $\left(\text{Find the range by solving for } x\colon x = \dfrac{2y^2 + 3}{y^2 - 1}.\right)$

y-axis: No, since the domain, $\{x \,|\, x \leq -3 \text{ or } x > 2\}$, is asymmetric about 0. Therefore, for example, although $(2.5, \sqrt{11})$ is on the graph, $(-2.5, \sqrt{11})$ cannot be on the graph because -2.5 is not in the domain. $\left(\text{Find the domain by observing that } y^2 \geq 0 \Rightarrow \dfrac{x+3}{x-2} \geq 0.\right)$

$(0,0)$: No. Recall that, if any two of these symmetries hold, then the third must hold. If there were symmetry about the origin, then there would be two symmetries but not the third, an impossibility.

$y = x$: No, since $(3, \sqrt{6})$ is on the graph and $(\sqrt{6}, 3)$ is not.

KEY DIAGRAM 12-8

If we look ahead to the discussion of asymptotes (KEYS 17-2 and 17-6), we can determine that $x = 2$, $y = 1$, and $y = -1$ are asymptotes and can draw a fairly accurate sketch of $y^2 = \dfrac{x+3}{x-2}$.

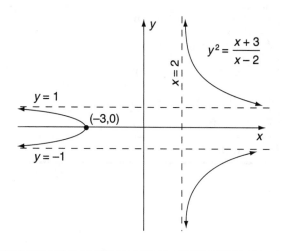

Theme 3 LIMITS AND CONTINUITY

When two variables are functionally related, it is reasonable to ask what is the behavior of one (dependent) variable as the other (independent) variable exhibits the specific behavior of "approaching" a particular number. When the dependent variable also approaches a particular number, we call that number a **limit**. Greek mathematicians 2000 years ago understood this concept and used it, among other things, to calculate the area of a circle. The concept of limit is very powerful, but the Greeks were limited in their advances because modern algebraic notation had not yet been invented. Limit is the fundamental concept of calculus. In fact, a simplistic definition of calculus would be a study of limits. It is in this theme that we begin our calculus adventure.

Key 13   An intuitive notion of limit

OVERVIEW *Every thought process of calculus involves limits. As theorems are proved and used to solve problems, the need to mention limits diminishes. However, without a complete understanding of the limit concept, these theorems become mechanical rules that, at best, are applied by rote. To master calculus, one must master limits. We start with an intuitive understanding.*

KEY EXAMPLE 13-1

Imagine that you are in a room, standing against a wall at one end. You start walking toward the opposite wall, each step being half the remaining distance to that wall. Do you ever reach the opposite wall? If not, exactly how close do you get to the wall?

Solution: Since every step is *half* the remaining distance, it is clear that you never reach the opposite wall. But how close do you get? Given an infinite amount of time, you will get ***closer than any distance you can name.*** If you want to be within a billionth of a millimeter of the wall, after taking enough steps you will! Mathematicians call the opposite wall the ***limit*** of your travels.

KEY DIAGRAM 13-2

Suppose that two variables, x and y, are functionally related. Consider the graph of $y = f(x)$. (See KEYS 13-3 and 13-4 for examples.)

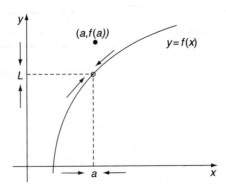

- For the graph above, as x gets closer and closer (closer than any distance we can name) to a (written as $x \to a$), the images of x are getting closer and closer (closer than any distance we can name) to L (written as $f(x) \to L$).
- We say, "The limit of the function $f(x)$ as x approaches a is equal to L." Notice that we do *not* say "approaches L." The limit *equals* L.
- In mathematical notation, we write $\lim\limits_{x \to a} f(x) = L$.
- IMPORTANT: What is happening *at* $x = a$ is not relevant to the limit itself. What is happening at $x = a$ is determined by the function's rule for finding the image of a. In this graph, $L \neq f(a)$, the limit does not equal the image.

The ability to evaluate limits is of paramount importance. Sometimes the limit is obvious, and other times it is not. In the latter case, algebraic manipulation is required.

KEY EXAMPLE 13-3

Evaluate: $\lim\limits_{x \to 3} x^2$.

Solution: As x approaches 3, your mind can envision x^2 (the images of x) approaching 9, that is, getting closer to 9 than any distance you can name. Therefore, $\lim\limits_{x \to 3} x^2 = 9$.

- In the preceding example, notice that you could have merely substituted 3 into the rule ($f(x) = x^2$) to calculate the image of 3 and obtained the correct limit. Although this happens frequently (see

KEY 18), mere substitution is not the way to understand limits. Calculate a limit correctly by investigating the behavior of the images as x approaches its limiting value. It pays not to take short-cuts, unless they are *proved* valid as theorems. Incorrect reasoning usually leads to incorrect results!

KEY EXAMPLE 13-4

Evaluate: $\lim\limits_{x \to 3} \dfrac{x^2 - 9}{x - 3}$.

Solution: $f(x) = \dfrac{x^2 - 9}{x - 3}$. Since 3 is not in the domain of f, you cannot substitute 3 for x, even if you wanted to find the image! But does the lack of an image imply that there is no limit? Not necessarily!

Notice that $\dfrac{x^2 - 9}{x - 3} = \dfrac{(x+3)\cancel{(x-3)}}{\cancel{x-3}} = x + 3$, for all $x \neq 3$. Recall that what is happening at $x = 3$ is not the issue, and reason that the behaviors of the functions $y = \dfrac{x^2 - 9}{x - 3}$ and $y = x + 3$ as x approaches 3 are identical. Therefore,

$$\lim_{x \to 3} \frac{x^2 - 9}{x - 3} = \lim_{x \to 3}(x + 3) = 6$$

Key 14 One-sided limits

OVERVIEW *When considering the limit of f(x) as x approaches a, we observe that x approaches a from two sides. It is possible that the limit as x approaches a from the right (a **right-hand limit**) may not equal the limit as x approaches a from the left (a **left-hand limit**).*

KEY NOTATION 14-1

A. A **right-hand limit** is denoted by $\lim\limits_{x \to a^+} f(x)$.

B. A **left-hand limit** is denoted by $\lim\limits_{x \to a^-} f(x)$.

- When calculating a right-hand limit, we consider only values of x **larger** than a as x approaches a.
- When calculating a left-hand limit, we consider only values of x **smaller** than a as x approaches a.

KEY THEOREM 14-2

The $\lim\limits_{x \to a} f(x)$ exists if and only if $\lim\limits_{x \to a^+} f(x) = \lim\limits_{x \to a^-} f(x)$.

- Therefore, if $\lim\limits_{x \to a^+} f(x) \neq \lim\limits_{x \to a^-} f(x)$, then $\lim\limits_{x \to a} f(x)$ does not exist.

KEY EXAMPLE 14-3

Evaluate: $\lim\limits_{x \to 3} x^2$ using one-sided limits.

Solution: $\lim\limits_{x \to 3^+} x^2 = 9$ and $\lim\limits_{x \to 3^-} x^2 = 9$. Therefore, $\lim\limits_{x \to 3} x^2 = 9$ (it exists).

KEY EXAMPLE 14-4

Evaluate: $\displaystyle\lim_{x \to 2} \frac{|x-2|}{x-2}$.

Solution: Use KEY 2-8; then

$$x > 2 \implies x - 2 > 0 \implies |x-2| = x - 2$$
$$x < 2 \implies x - 2 < 0 \implies |x-2| = -(x-2)$$

Therefore,

$$\lim_{x \to 2^+} \frac{|x-2|}{x-2} = \lim_{x \to 2^+} \frac{x-2}{x-2} = 1 \quad \text{and} \quad \lim_{x \to 2^-} \frac{|x-2|}{x-2} = \lim_{x \to 2^-} \frac{-(x-2)}{x-2} = -1$$

Since the right-hand limit does not equal the left-hand limit, $\displaystyle\lim_{x \to 2} \frac{|x-2|}{x-2}$ does not exist.

KEY DIAGRAM 14-5

Consider the graph of $y = \dfrac{|x-2|}{x-2}$.

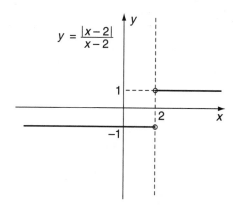

$$y = \frac{|x-2|}{x-2}$$

- It should be obvious from the graph that

$$\lim_{x \to 2^+} \frac{|x-2|}{x-2} = 1 \quad \text{and} \quad \lim_{x \to 2^-} \frac{|x-2|}{x-2} = -1$$

- **WARNING:** Do not assume that, if a function f is defined piecewise (KEY 2-7), with different rules for calculating images on the left and the right of a, then $\lim_{x \to a} f(x)$ does not exist. Instead, calculate the left-hand and right-hand limits for those values of a and then make a judgment.

Key 15 Definition of a limit

OVERVIEW *The intuitive idea that a limit is a number that some variable is "approaching" or "getting closer and closer to" is, at best, imprecise and, at worst, misleading. Mathematicians have a rigorous, precise definition of limit. Although it is important to understand, as completely as possible, this fundamental concept of calculus, many calculus courses do not require this rigor. Each student must individually decide the depth of knowledge that is desirable.*

KEY DEFINITION 15-1

$\lim_{x \to a} f(x) = L$ if and only if, for each real number $\varepsilon > 0$, there must exist a real number $\delta > 0$ such that

$$0 < |x - a| < \delta \implies |f(x) - L| < \varepsilon$$

In words, for each x closer than δ to a, its image, $f(x)$, is closer than ε to the limit L. See KEY 15-2.

- WARNING: There is nothing in the definition that prohibits a function from *reaching* its limit; after all, $0 < \varepsilon$!

KEY DIAGRAM 15-2

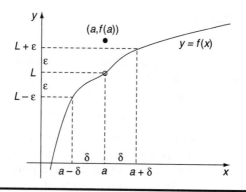

- Do not worry about the symbols ε and δ. They are the Greek lower-case letters epsilon and delta, respectively, and are universally used in this situation.
- From KEY 2-11, since $\delta > 0$,

$$|x - a| < \delta \quad \Leftrightarrow \quad a - \delta < x < a + \delta$$

In other words, x is within distance δ from a. Similarly, since $\varepsilon > 0$,

$$|f(x) - L| < \varepsilon \quad \Leftrightarrow \quad L - \varepsilon < f(x) < L + \varepsilon.$$

In words, the images, $f(x)$, of x are within distance ε from the limit L.
- The definition (KEY 15-1) states that $\lim_{x \to a} f(x) = L$ means that, by staying close enough (δ) to a, **all** of the images are within **any** distance (ε) from the limit (L) you can name.
- In the definition of limit, $0 < |x - a|$ means $x \neq a$. Recall that the value of $f(a)$ has nothing to do with the limit.
- The following comments help to clarify the relationship between ε and δ. Examination of the diagram indicates that (1) how close you want the images to be to the limit will dictate how close you need to stay to a and, in general, as ε is chosen to be smaller, δ will probably need to be smaller as well; (2) if a particular δ "works" for a particular ε, then any smaller δ will work for the same ε (δ is not unique); and (3) if a particular δ works for a particular ε, then that same δ will work for any larger ε.

KEY EXAMPLE 15-3

You can "see" that $\lim_{x \to 3}(2x + 1) = 7$. Use the definition of limit to *prove* the limit is 7.

Solution: By definition, you need to show that for any $\varepsilon > 0$ (no matter how close you want to be to the limit), there exists a $\delta > 0$ (it must be possible to stay close enough to 3) such that

$$0 < |x - 3| < \delta \quad \Rightarrow \quad |(2x + 1) - 7| < \varepsilon$$

(staying that close to 3 guarantees that *every* image is within the required distance from 7—the limit). To accomplish this, for every possible $\varepsilon > 0$, you must find the $\delta > 0$ that will work and then *prove* it. To find δ, work backward (the "aside").

THE ASIDE:

$$|(2x-1)-7|<\varepsilon \quad \Rightarrow \quad |2x-6|<\varepsilon$$
$$\Rightarrow \quad |2\|x-3|<\varepsilon \quad \text{(KEY 2-9D)}$$
$$\Rightarrow \quad |x-3|<\frac{\varepsilon}{|2|}=\frac{\varepsilon}{2}$$

This suggests that $\delta=\dfrac{\varepsilon}{2}$.

THE PROOF: For any real number $\varepsilon>0$, choose $\delta=\dfrac{\varepsilon}{2}$. Then:

$$|x-3|<\delta \quad \Rightarrow \quad |x-3|<\frac{\varepsilon}{2} \Rightarrow |x-3|<\frac{\varepsilon}{|2|}$$
$$\Rightarrow \quad |2\|x-3|<\varepsilon \Rightarrow |2x-6|<\varepsilon \quad \text{(KEY 2-9D)}$$
$$\Rightarrow \quad |(2x+1)-7|<\varepsilon$$

In this particular problem, could you have predicted that $\delta=\dfrac{\varepsilon}{2}$?

KEY DIAGRAM 15-4

Consider the graph of $y=2x+1$.

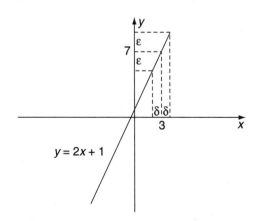

A. Since the slope of the line $y = 2x + 1$ is 2 and $m = \dfrac{\Delta y}{\Delta x}$, the change in y is twice the change in x, or the change in x is one-half the change in y. Since δ represents how much x can change and ε represents how much y can change, it follows that $\delta = \dfrac{\varepsilon}{2}$.

B. We can generalize: If we are proving that our evaluation of $\lim\limits_{x \to a}(mx + b)$ is accurate using an "epsilon-delta proof," then it will turn out that $\delta = \dfrac{\varepsilon}{|m|}$. The reason for the absolute value is that, if the slope is negative, delta must still be positive.

Key 16   Limit theorems

OVERVIEW *So far, the only method of rigorously determining the limit of a function is by observation and use of
the definition (an epsilon-delta proof); these proofs can get
very complicated. In this section, we discuss theorems that
permit fast but definitive evaluation of limits.*

KEY THEOREM 16-1

The limit of a function as $x \to a$, if it exists, is unique.

- In other words, $\lim_{x \to a} f(x)$ cannot have two different values.

KEY THEOREM 16-2

If $f(x) = g(x)$ for $0 < |x - a| < \delta$, then $\lim_{x \to a} f(x) = \lim_{x \to a} g(x)$.

- "If, for *all* x close enough to a but not equal to a (recall that what is
happening at $x = a$ is irrelevant to the limit), two functions are identical, then their limits must be equal." This theorem was already used
in the solution of KEY 13-4.

KEY THEOREM 16-3

If $f(x) = c$, then $\lim_{x \to a} f(x) = c$.

- "No matter which number x is approaching, the limit of a constant
function is that constant."

KEY THEOREM 16-4

$$\lim_{x \to a} x = a$$

- "For the function $f(x) = x$, its limit as x approaches any number a is that number, a."

KEY THEOREM 16-5

$$\lim_{x \to a} x^n = a^n$$

Example: $\lim_{x \to 3} x^4 = 3^4 = 81$

- Until this theorem was introduced, the verification of this limit would have been a complicated process using the definition of a limit (an epsilon-delta proof). This theorem not only helps to calculate the limit, but also *guarantees* the result. This observation applies to all theorems in this section.

KEY THEOREM 16-6

$$\lim_{x \to a} cf(x) = c \lim_{x \to a} f(x)$$

- "The limit of a constant times a function equals the constant times the limit of the function."
- **NOTE: This theorem and those that follow assume that all limits exist. If any limit fails to exist, then further analysis is necessary.**

Example: $\lim_{x \to 3} 5x^4 = 5 \lim_{x \to 3} x^4 = 5 \cdot 3^4 = (5)(81) = 405$

KEY THEOREM 16-7

$$\lim_{x \to a} [f(x) + g(x)] = \lim_{x \to a} f(x) + \lim_{x \to a} g(x)$$

- "The limit of a sum equals the sum of the respective limits."

Example: $\lim_{x\to 2}(x^3 + x^2) = \lim_{x\to 2} x^3 + \lim_{x\to 2} x^2 = 2^3 + 2^2 = 8 + 4 = 12$

KEY THEOREM 16-8

$$\lim_{x\to a}[f(x) - g(x)] = \lim_{x\to a} f(x) - \lim_{x\to a} g(x)$$

- "The limit of a difference equals the difference of the respective limits."

Examples: $\lim_{x\to 2}(x^3 - x^4) = \lim_{x\to 2} x^3 - \lim_{x\to 2} x^4 = 2^3 - 2^4 = 8 - 16 = -8$

$$\begin{aligned}
\lim_{x\to 2}(x^4 - 2x^3 - 5x + 6) &= \lim_{x\to 2} x^4 - \lim_{x\to 2} 2x^3 - \lim_{x\to 2} 5x + \lim_{x\to 2} 6 \\
&= \lim_{x\to 2} x^4 - 2\lim_{x\to 2} x^3 - 5\lim_{x\to 2} x + \lim_{x\to 2} 6 \\
&= 2^4 \quad - 2\cdot 2^3 \quad - 5\cdot 2 \quad + 6 \\
&= 16 \quad -16 \quad -10 \quad +6 = -4
\end{aligned}$$

KEY THEOREM 16-9

$$\lim_{x\to a}[f(x)g(x)] = \lim_{x\to a} f(x) \cdot \lim_{x\to a} g(x)$$

- "The limit of a product equals the product of the respective limits."

Example: $\lim_{x\to -1}(x^3 + 4)(x^2 - 8) = \lim_{x\to -1}(x^3 + 4)\lim_{x\to -1}(x^2 - 8) = (3)(-7) = -21$

KEY THEOREM 16-10

$$\lim_{x\to a}\frac{f(x)}{g(x)} = \frac{\lim_{x\to a} f(x)}{\lim_{x\to a} g(x)}, \text{ if } \lim_{x\to a} g(x) \neq 0$$

- "The limit of a quotient equals the quotient of the respective limits, provided that the limit of the denominator is not 0." If the limit of the denominator *is* 0, refer to KEYS 17-1E and 17-1F.

Example: $\lim\limits_{x \to -2} \dfrac{x^2 - 10}{x^3 + x} = \dfrac{\lim\limits_{x \to -2}(x^2 - 10)}{\lim\limits_{x \to -2}(x^3 + x)} = \dfrac{-6}{-10} = \dfrac{3}{5}$

KEY THEOREM 16-11

$$\lim_{x \to a} \sqrt[n]{f(x)} = \sqrt[n]{\lim_{x \to a} f(x)}$$

- "The limit of an *n*th root of a function equals the *n*th root of the limit of the function," provided that the result is a real number (not imaginary).

Example: $\lim\limits_{x \to 2} \sqrt[3]{x^4 + 7} = \sqrt[3]{\lim\limits_{x \to 2}(x^4 + 7)} = \sqrt[3]{23}$

KEY THEOREM 16-12

If, for some δ,

$$0 < |x - a| < \delta \quad \Rightarrow \quad f(x) \le g(x)$$

then $\lim\limits_{x \to a} f(x) \le \lim\limits_{x \to a} g(x)$.

KEY THEOREM 16-13

If, for some δ,

$$0 < |x - a| < \delta \quad \Rightarrow \quad f(x) \le g(x) \le h(x)$$

and

$$\lim_{x \to a} f(x) = \lim_{x \to a} h(x) = L$$

then $\lim\limits_{x \to a} g(x) = L$.

- "Of three functions, if, for *all* x close enough to *a* but not equal to *a,* the images of one function are between the images of the other two functions, and, furthermore, the limit of the smaller function is equal to the limit of the larger function, then the 'middle' function must have the same limit."

- This theorem is frequently called a "comparison" or "squeeze" theorem.

KEY OBSERVATION 16-14

All of the limit theorems apply to the different types of limits, including one-sided limits and infinite limits (again assuming the limits exist).

KEY EXAMPLE 16-15

Evaluate: $\lim\limits_{x \to \infty} \dfrac{\sin x}{x}$.

Solution: You cannot use KEY 16-10 because the numerator has no limit (it oscillates between 1 and −1 and does not approach a particular number) and the denominator has no limit (it grows larger without bound). However, you may reason as follows:

$-1 \le \sin x \le 1$ (the range of the sine function)

$\dfrac{-1}{x} \le \dfrac{\sin x}{x} \le \dfrac{1}{x}$ (for all $x > 0$)

$\lim\limits_{x \to \infty} \dfrac{-1}{x} \le \lim\limits_{x \to \infty} \dfrac{\sin x}{x} \le \lim\limits_{x \to \infty} \dfrac{1}{x}$

$\lim\limits_{x \to \infty} \dfrac{-1}{x} = 0$ and $\lim\limits_{x \to \infty} \dfrac{1}{x} = 0$ $\Rightarrow$ $\lim\limits_{x \to \infty} \dfrac{\sin x}{x} = 0$ (KEY 16-13)

Key 17   Infinite limits and asymptotes

OVERVIEW *Thus far, we have confined the discussion to limits of functions as the independent variable approaches a finite number. Is it reasonable to attempt to evaluate the limit of a function as the independent variable grows infinitely large? In addition, what is the significance of images growing infinitely large as the independent variable approaches a finite number? These questions lead to two different types of infinite limits.*

KEY OBSERVATIONS 17-1

When evaluating the limit of a fraction:

A. If the numerator remains finite and the denominator grows infinitely large, then the limit of the fraction is 0.

B. If the numerator grows infinitely large and the denominator remains finite, then the fraction grows infinitely large (positively or negatively).

C. If the numerator and denominator *both* grow infinitely large, then no conclusion can be made; further analysis is necessary.

D. If the limit of the numerator *is* 0 and the limit of the denominator is *not* 0, then the limit of the fraction is 0.

E. If the limit of the numerator is *not* 0 and the limit of the denominator *is* 0, then the fraction grows infinitely large (positively or negatively).

F. If the limits of the numerator and denominator are *both* 0, then no conclusion can be made; further analysis is necessary.

KEY EXAMPLE 17-2

Evaluate: $\lim\limits_{x\to\infty}\dfrac{10,000}{x}$.

Solution: By KEY 17-1A, $\lim\limits_{x\to\infty}\dfrac{10,000}{x} = 0$. Also, $\lim\limits_{x\to-\infty}\dfrac{10,000}{x} = 0$.

- In this case, the graph of $y = \dfrac{10{,}000}{x}$ has a **horizontal asymptote**, $y = 0$.
- In general, if, as $x \to \pm\infty$, $y \to a$ (a finite number), then the graph of $y = f(x)$ has a horizontal asymptote, $y = a$.

KEY EXAMPLE 17-3

Evaluate: $\displaystyle\lim_{x\to\infty} \dfrac{x}{10{,}000}$.

Solution: By KEY 17-1B, $\displaystyle\lim_{x\to\infty} \dfrac{x}{10{,}000} = \infty$.

- In this case, the limit does not exist, but we write that the limit equals "infinity" to indicate that there is no limit because the images are growing infinitely large. Also, $\displaystyle\lim_{x\to-\infty} \dfrac{x}{10{,}000} = -\infty$.

KEY EXAMPLE 17-4

Evaluate: $\displaystyle\lim_{x\to\infty} \dfrac{5x^4 + 2x + 1}{3x^4 + 7x^2}$.

Solution: Since both the numerator and the denominator are growing infinitely large, further analysis is necessary. **When the numerator and denominator are polynomials, dividing both by x raised to the highest power in the expression helps.**

$$\lim_{x\to\infty} \frac{5x^4 + 2x + 1}{3x^4 + 7x^2} = \lim_{x\to\infty} \frac{\dfrac{5x^4 + 2x + 1}{x^4}}{\dfrac{3x^4 + 7x^2}{x^4}} = \lim_{x\to\infty} \frac{\dfrac{5x^4}{x^4} + \dfrac{2x}{x^4} + \dfrac{1}{x^4}}{\dfrac{3x^4}{x^4} + \dfrac{7x^2}{x^4}}$$

$$= \lim_{x\to\infty} \frac{5 + \dfrac{2}{x^3} + \dfrac{1}{x^4}}{3 + \dfrac{7}{x^2}} = \frac{5 + 0 + 0}{3 + 0} \quad \text{(KEY 17-1A)}$$

$$= \frac{5}{3}$$

- Notice that the symbol "lim" is written until the process of taking the limit is performed, after which the symbol is omitted.
- Using the technique of KEY 17-4, you should verify the following two limits:

$$\lim_{x \to \infty} \frac{5x^3 + 2x + 1}{3x^4 + 7x^2} = 0 \quad \text{and} \quad \lim_{x \to \infty} \frac{5x^4 + 2x + 1}{3x^3 + 7x^2} = \infty$$

- From the preceding three limits, you should understand why mathematicians call $\dfrac{\infty}{\infty}$ an **indeterminate form**.

KEY EXAMPLE 17-5

Evaluate: $\lim\limits_{x \to -5} \dfrac{x + 5}{x^2 - 3x + 4}$.

Solution: Since the numerator is approaching 0 and the denominator is not approaching 0, then, by KEY 17-1D, $\lim\limits_{x \to -5} \dfrac{x + 5}{x^2 - 3x + 4} = 0$.

KEY EXAMPLE 17-6

Evaluate: $\lim\limits_{x \to -5} \dfrac{x^2 - 3x + 4}{x + 5}$.

Solution: By KEY 17-1E, $\lim\limits_{x \to -5} \dfrac{x^2 - 3x + 4}{x + 5} = \pm\infty$ (different results

approaching from the left and from the right—in each case, there is no limit). The result becomes clearer if you consider the left-hand limit and the right-hand limit. If x is very close to -5 and $x < -5$, then the numerator is positive and the denominator is negative. Therefore, $\lim\limits_{x \to -5^-} \dfrac{x^2 - 3x + 4}{x + 5} = -\infty$ and the images are getting infinitely larger negatively ("without bound"). If x is very close to -5 and $x > -5$, then the numerator is positive and the denominator is positive. Therefore, $\lim\limits_{x \to -5^+} \dfrac{x^2 - 3x + 4}{x + 5} = +\infty$ and the images are getting infinitely larger positively ("without bound").

- In this case, the graph of $y = \dfrac{x^2 - 3x + 4}{x + 5}$ has a **vertical asymptote**, $x = -5$.
- In general, if, as $x \to a$ (from the right, from the left, or both), $y \to \pm\infty$, then the graph of $y = f(x)$ has a vertical asymptote, $x = a$.
- In searching for vertical asymptotes, if $x = f(y)$, reverse the order of the inquiry: let $y \to \pm\infty$, and determine whether x approaches a finite number a.

KEY EXAMPLES 17-7

By KEY 17-1F, when the limits of the numerator and denominator are both 0, further analysis is necessary.

A. $\displaystyle\lim_{x \to 2} \frac{x^2 - 4}{x^2 - x - 1} = \lim_{x \to 2} \frac{(x+2)(x-2)}{(x+1)(x-2)} = \lim_{x \to 2} \frac{x+2}{x+1} = \frac{4}{3}$

B. $\displaystyle\lim_{x \to 2} \frac{x^2 - 4x + 4}{x^2 - 4} = \lim_{x \to 2} \frac{(x-2)(x-2)}{(x+2)(x-2)} = \lim_{x \to 2} \frac{x-2}{x+2} = 0$ (KEY 17-1D)

C. $\displaystyle\lim_{x \to 2} \frac{x^2 - 4}{x^2 - 4x + 4} = \lim_{x \to 2} \frac{(x+2)(x-2)}{(x-2)(x-2)} = \lim_{x \to 2} \frac{x+2}{x-2} = \pm\infty$ (KEY 17-1D)

$$\lim_{x \to 2^-} \frac{x+2}{x-2} = -\infty \quad \text{and} \quad \lim_{x \to 2^+} \frac{x+2}{x-2} = +\infty$$

Also, the graph of $y = \dfrac{x+2}{x-2}$ has a vertical asymptote, $x = 2$. (See KEY 17-8.)

- From the preceding examples, it should be clear why $\dfrac{0}{0}$ is another **indeterminate form**.

KEY EXAMPLE 17-8

Determine the asymptotes of the graph of $y = \dfrac{x+2}{x-2}$.

Solution: First examine the behavior of the graph near $x = 2$ because this behavior causes the denominator to equal 0. Since

$$\lim_{x \to 2^-} \frac{x+2}{x-2} - \infty \quad \text{and} \quad \lim_{x \to 2^+} \frac{x+2}{x-2} = +\infty$$

$x = 2$ is a vertical asymptote. Also, since

$$\lim_{x\to\infty}\frac{x+2}{x-2}=\lim_{x\to\infty}\frac{\dfrac{x+2}{x}}{\dfrac{x-2}{x}}=\lim_{x\to\infty}\frac{1+\dfrac{2}{x}}{1-\dfrac{2}{x}}=\frac{1+0}{1-0}=1$$

$y = 1$ is a horizontal asymptote, as indicated on the following graph.

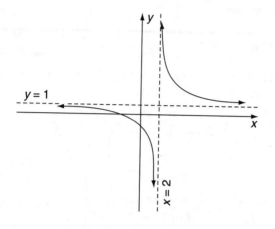

Key 18   Continuity

OVERVIEW *Sometimes the limit of a function as **x** approaches **a** is equal to the image of **a**, and sometimes the limit is not equal to the image. Wherever the limit equals the image, mathematicians say that the function is continuous.*

KEY DEFINITION 18-1

A function, f, is **continuous at $x = a$** if and only if

(1) $\lim\limits_{x \to a} f(x)$ exists,

(2) $f(a)$ exists, and

(3) $\lim\limits_{x \to a} f(x) = f(a)$.

- Continuity at $x = a$ means "the limit equals the image."
- Continuity, like limit, is a "pointwise property" in that a function may be continuous at some numbers in its domain and discontinuous at others.
- Continuity of a function at $x = a$ essentially means that the graph of the function is "unbroken" at $x = a$.
- The question of continuity at $x = a$ is not raised if a is not in the domain. For example, with respect to the function $f(x) = \sqrt{x}$, continuity is not relevant at $x = -4$ because -4 is not "close" to the domain and, therefore, x cannot even approach -4.
- According to the definition (KEY 18-1), the function $f(x) = \sqrt{x}$ is discontinuous at $x = 0$ because $\lim\limits_{x \to 0^-} \sqrt{x}$ fails to exist and, therefore, $\lim\limits_{x \to 0} \sqrt{x}$ fails to exist. However, the right-hand limit, $\lim\limits_{x \to 0^+} \sqrt{x}$, *does* equal the image of 0. Mathematicians call this **right-hand continuity**.

KEY DEFINITIONS 18-2

A. **Left-hand continuity at $x = a$** means $\lim\limits_{x \to a^-} f(x) = f(a)$.

B. **Right-hand continuity at $x = a$** means $\lim\limits_{x \to a^+} f(x) = f(a)$.

KEY DEFINITION 18-3

A function is **continuous on an interval** (KEY 2-4) if and only if the function is continuous at every point in the interval except at the endpoints (if included in the interval), at which the function must have the appropriate one-sided continuity.

KEY EXAMPLES 18-4

Discuss continuity on the closed interval [0, 1] for each of the following functions:

A. $y = \sqrt{x}$: This function *is* continuous on the interval. Note that there is right-hand continuity at $x = 0$ and left-hand continuity at $x = 1$.

B. $y = \dfrac{1}{x}$: This function is discontinuous at $x = 0$ since neither $f(0)$ nor $\lim\limits_{x \to 0^+} \dfrac{1}{x}$ exists. However, $y = \dfrac{1}{x}$ *is* continuous on the interval (0, 1].

KEY DIAGRAMS 18-5

Each of the following graphs is discontinuous at $x = a$ for the stated reasons.

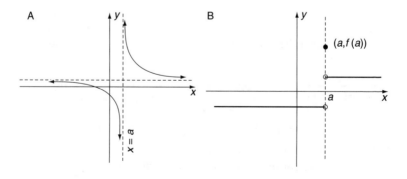

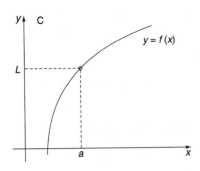 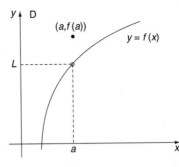

A. The graph is discontinuous at $x = a$ because both $\lim\limits_{x \to a} f(x)$ and $f(a)$ do not exist. This particular graph has what is sometimes called an "**infinite discontinuity**."

B. The graph is discontinuous at $x = a$ because $\lim\limits_{x \to a} f(x)$ does not exist (the left-hand limit does not equal the right-hand limit). This particular graph has what is sometimes called a "**finite (or jump) discontinuity**."

C. The graph is discontinuous at $x = a$ because, although $\lim\limits_{x \to a} f(x)$ does exist, $f(a)$ does not exist. This situation is called a **removable discontinuity** because it is possible to define a value for $f(a)$ that would "fill the hole" in the graph.

D. The graph is discontinuous at $x = a$ because $\lim\limits_{x \to a} f(x) \neq f(a)$. This is also a removable discontinuity because $f(a)$ can be redefined to make the graph continuous at $x = a$.

KEY EXAMPLE 18-6

Let $f(x) = \dfrac{x^2 - 9}{x - 3}, x \neq 3$. What value should be assigned to $f(3)$ so that the function f is continuous at $x = 3$?

Solution: Because $\lim\limits_{x \to 3} \dfrac{x^2 - 9}{x - 3} = \lim\limits_{x \to 3}(x + 3) = 6$ and, for continuity, the image must equal the limit, define $f(3) = 6$.

- The graph of $y = \dfrac{x^2 - 9}{x - 3}$ is the graph of $y = x + 3$ with a hole (a removable discontinuity) at point (3,6).

KEY THEOREMS 18-7

If two functions, f and g, are each continuous at $x = a$, then:

A. $f + g$ is continuous at $x = a$.
B. $f - g$ is continuous at $x = a$.
C. $f \cdot g$ is continuous at $x = a$.
D. $\dfrac{f}{g}$ is continuous at $x = a$, provided that $g(a) \neq 0$.

KEY THEOREMS 18-8

A. Every polynomial function is continuous at every real number.
B. A rational function (the ratio of two polynomials) is continuous in its domain, that is, for all reals except where the denominator is 0.

KEY EXAMPLE 18-9

Evaluate $\lim\limits_{x \to 2}(x^4 - 2x^3 - 5x + 6)$, and prove your result.

Solution: Since $f(x) = x^4 - 2x^3 - 5x + 6$ is a polynomial function, by KEY 18-8A, it is continuous at $x = 2$. Therefore,

$$\lim\limits_{x \to 2}(x^4 - 2x^3 - 5x + 6) = f(2) = -4$$

No other proof is necessary.

KEY THEOREM 18-10

If f is continuous at $x = a$ and g is continuous at $x = f(a)$, then the *composite* function $(g \circ f)(x) = g(f(x))$ is continuous at $x = a$.

KEY OBSERVATION 18-11

Using KEY 16-11, we can conclude that all "root functions" are continuous in their domains. For example, $y = \sqrt{x-1}$ is continuous for all $x \geq 1$ and $y = \sqrt[3]{x}$ is continuous for all real numbers.

KEY EXAMPLE 18-12

Show that the absolute-value function is continuous for all real numbers.

Solution: Recall that $|x| = \sqrt{x^2}$ (KEY 2-9), which is a composite function, $g \circ f$, with $f(x) = x^2$ and $g(x) = \sqrt{x}$. Also, f is a polynomial function that is continuous for all reals (KEY 18-8), g is a root function that is continuous for all $x \geq 0$, and $x^2 \geq 0$ for all reals. Therefore, the composite function $g(f(x)) = \sqrt{x^2} = |x|$ is continuous for all reals.

KEY EXAMPLE 18-13

Discuss the continuity of f at all integers between -2 and 3 inclusive, where

$$f(x) = \begin{cases} 3x - 5, & x < -1 \\ x^2, & x = -1 \\ \dfrac{8}{x}, & -1 < x < 2 \\ 2x + 4, & x \geq 2 \end{cases}$$

Solutions: f is:

Continuous at $x = -2$. The rule is a polynomial function.

Discontinuous at $x = -1$. Since the left-hand and right-hand limits are equal, $\lim\limits_{x \to -1} f(x) = -8$. But $f(-1) = 1$, so the limit does not equal the image. This is a removable discontinuity.

Discontinuous at $x = 0$. Neither the limit nor the image exists. This is an infinite discontinuity.

Continuous at $x = 1$. The rule is a rational function, and 1 is in its domain.

Discontinuous at $x = 2$. Since the left-hand limit (4) does not equal the right-hand limit (8), the limit does not exist. This is a finite discontinuity.

Continuous at $x = 3$. The rule is a polynomial function.

KEY EXAMPLE 18-14

Find the value(s) of p and q such that $g(x)$ is continuous at $x = 1$, if

$$g(x) = \begin{cases} p^2qx, & x < 1 \\ -p, & x = 1 \\ 3px + q, & x > 1 \end{cases}$$

Solution: To have continuity at $x = 1$, the left-hand limit and the right-hand limit both have to equal $g(1)$.

$$\lim_{x \to 1^-} g(x) = p^2q \quad \text{and} \quad \lim_{x \to 1^+} g(x) = 3p + q$$

Therefore, $3p + q = -p$ and $p^2q = -p$. Also, since $q = -4p$, then $p^2(-4p) = -p$ and $4p^3 = p$.

$$4p^3 = p \quad \Rightarrow \quad 4p^3 - p = 0 \quad \Rightarrow \quad p(2p-1)(2p+1) = 0$$

$$\Rightarrow \quad p = 0, q = 0 \text{ or } p = \frac{1}{2}, q = -2 \text{ or } p = -\frac{1}{2}, q = 2$$

Theme 4 THE DERIVATIVE

*F*or centuries a classic problem in mathematics (and science) was how to calculate **instantaneous rate of change**. This ability was especially important when attempting to define the precise motion of moving objects on Earth and the motions of the planets. Sir Isaac Newton (1642–1727) and Gottfried Wilhelm Leibniz (1646–1716) solved this problem by using a special limit (called *fluxion* by Newton) that is now termed a **derivative**.

Key 19 Average rate of change

OVERVIEW *When calculating the rate of change of a (dependent) variable, we must compare it to another **related** (independent) variable. The unit of measure for average rate of change is change in the dependent variable per **unit change** in the independent variable, as in miles per (1) hour. In mathematics, the change in a variable v is usually indicated by Δv.*

Average rate of change: To calculate the average rate of change of one variable **with respect to** (as compared to) a second variable, we divide the net change of the first variable by the change in the second variable. (See KEY 19-1.)

KEY DEFINITION 19-1

If the position, s, of an object is a function of time, $s = f(t)$, then the average rate of change of position with respect to time over an interval of time $t_1 \leq t \leq t_2$ is

$$\bar{v} = \frac{\Delta s}{\Delta t} = \frac{f(t_2) - f(t_1)}{t_2 - t_1}$$

KEY NOTE 19-2

In Key 19-1 "with respect to" is a very important phrase. For example, the speedometer reports the speed of a car with respect to the ground (e.g., $50\,\text{mi/hr}$), but the speed of the car with respect to an overtaken truck may be different (e.g., $10\,\text{mi/hr}$). However, the phrase may be omitted if, from the context, the independent variable is obvious. For example, we often say "the change in y" rather than "the change in y with respect to x."

KEY EXAMPLE 19-3

An object is moving along a path in such a way that its position, s (in meters), from some reference point is given by $s = f(t) = t^3 + 2t + 3$, where time, t, is in seconds. What is the average velocity, $\bar{v}$ (rate of change in position with respect to time), for $1 \le t \le 4$?

Solution: Use KEY 19-1 with $t_1 = 1$ and $t_2 = 4$; then

$$\bar{v} = \frac{\Delta s}{\Delta t} = \frac{f(4) - f(1)}{4 - 1} = \frac{75 - 6}{3} = 23 \text{ m/s}$$

KEY EXAMPLE 19-4

If $y = f(x) = x^2 - 20x$, what is the average change in y with respect to x over the interval $3 \le x \le 7$? NOTE: There are no explicit units in this problem.

Solution: Use KEY 19-1; then

$$\frac{\Delta y}{\Delta x} = \frac{\Delta f}{\Delta x} = \frac{f(7) - f(3)}{7 - 3} = \frac{(-91) - (-51)}{4} = -10$$

Geometric interpretation: Consider the graph of any function $y = f(x)$. What would be a geometric representation of the average rate of change of y (with respect to x) over the interval $x_1 \le x \le x_2$?

KEY DIAGRAM 19-5

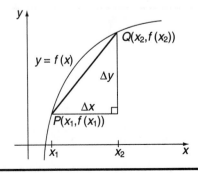

Here, $\dfrac{\Delta y}{\Delta x} = \dfrac{f(x_2) - f(x_1)}{x_2 - x_1} = m_{\overline{PQ}}$. The slope of $\overline{PQ}$ is the average rate of change of y.

NOTE: $\Delta y = f(x_2) - f(x_1) > 0$ and $\Delta x = x_2 - x_1 > 0$. Therefore, $m_{\overline{PQ}} = \dfrac{\Delta y}{\Delta x} > 0$.

KEY EXAMPLE 19-6

Draw a diagram in which the average rate of change of y (with respect to x) is *negative* over the interval $[x_1, x_2]$, that is, $x_1 \le x \le x_2$.

Solution:

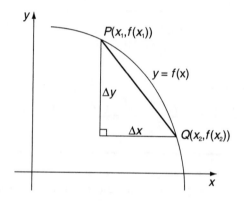

NOTE: Here, $\Delta y = f(x_2) - f(x_1) < 0$ while $\Delta x = x_2 - x_1 > 0$. Therefore, $\dfrac{\Delta y}{\Delta x} < 0$.

Key 20 Instantaneous rate of change

OVERVIEW *The instantaneous rate of change of a dependent variable with respect to an independent variable can be thought of as an average rate of change over "a very, very small interval"—an "interval" that has length 0. Since division by 0 is undefined, we use limits to define instantaneous rate of change.*

Instantaneous rate of change: If $y = f(x)$, we can obtain an approximation to the *instantaneous* rate of change of y with respect to x at $x = x_1$ by calculating the *average* rate of change over the interval $[x_1, x_1 + \Delta x]$ for some value of Δx: $\dfrac{f(x_1 + \Delta x) - f(x_1)}{\Delta x}$. We can always improve the approximation by choosing Δx closer to 0. This reasoning leads us to KEY 20-1.

KEY DEFINITION 20-1

At $x = x_1$

$$\text{Instantaneous change of } y \text{ with respect to } x = \lim_{\Delta x \to 0} \frac{f(x_1 + \Delta x) - f(x_1)}{\Delta x}$$

KEY EXAMPLE 20-2

An object is moving along a path in such a way that its position s, in meters, from some reference point is given by $s = f(t) = t^2 + 3t + 4$, where time, t, is in seconds. What is the instantaneous velocity, v (rate of change in position with respect to time), when $t = 2$ seconds?

Solution: Use KEY 20-1 with $x = t$ and $x_1 = 2$; then

$$v(2) = \lim_{\Delta t \to 0} \frac{f(2 + \Delta t) - f(2)}{\Delta t}$$

$$= \lim_{\Delta t \to 0} \frac{\left[(2 + \Delta t)^2 + 3(2 + \Delta t) + 4\right] - \left[2^2 + 3(2) + 4\right]}{\Delta t}$$

$$= \lim_{\Delta t \to 0} \frac{\left[4 + 4(\Delta t) + (\Delta t)^2 + 6 + 3(\Delta t) + 4 \right] - 14}{\Delta t}$$

$$= \lim_{\Delta t \to 0} \frac{4(\Delta t) + (\Delta t)^2 + 3(\Delta t)}{\Delta t}$$

$$= \lim_{\Delta t \to 0} [4 + (\Delta t) + 3] = 7$$

Therefore, the instantaneous velocity at $x = 2$ is 7 m/s.

KEY EXAMPLE 20-3

(alternative generalized solution to KEY 20-2)

An object is moving along a path in such a way that its position, s, in meters, from some reference point is given by $s = f(t) = t^2 + 3t + 4$, where time, t, is in seconds. What is the instantaneous velocity, v (rate of change in position with respect to time), when $t = 2$ seconds?

Solution: Use KEY 20-1 with $x_1 = t_1$; then

$$v(t_1) = \lim_{\Delta t \to 0} \frac{f(t_1 + \Delta t) - f(t_1)}{\Delta t}$$

$$= \lim_{\Delta t \to 0} \frac{\left((t_1 + \Delta t)^2 + 3(t_1 + \Delta t) + 4 \right) - (t_1^2 + 3t_1 + 4)}{\Delta t}$$

$$= \lim_{\Delta t \to 0} \frac{t_1^2 + 2t_1(\Delta t) + (\Delta t)^2 + 3t_1 + 3(\Delta t) + 4 - t_1^2 - 3t_1 - 4}{\Delta t}$$

$$= \lim_{\Delta t \to 0} \frac{2t_1(\Delta t) + (\Delta t)^2 + 3(\Delta t)}{\Delta t}$$

$$= \lim_{\Delta t \to 0} [2t_1 + (\Delta t) + 3]$$

$$= 2t_1 + 3$$

The instantaneous velocity at $t = t_1$ is $v(t_1) = 2t_1 + 3$. Hence, the instantaneous velocity at $t_1 = 2$ is 7 m/s.

- The alternative generalized solution is better because, with the same algebra, we derived an instantaneous rate of change *function* ($2t_1 + 3$) with which instantaneous velocity at *any time* can be evaluated. When a *general* solution requires no greater effort to find than a *particular* solution, we shall find the general solution first since it contains more information and always includes the particular solution in question.

- In the above solution, we shall call t_1 a **constant variable** (KEY 6-5), representing *any* real number that is treated as a *fixed* number in the discussion. Calculus textbooks usually drop the subscript (using t) because the context is clear. Henceforth, we shall do likewise.

- When calculating instantaneous rate of change of a dependent variable with respect to an independent variable, we have used as the change in the independent variable symbols such as Δx and Δt. Over the past few decades, however, it has become customary to use h, as in

$$\lim_{h \to 0} \frac{f(x+h) - f(x)}{h} \quad \text{as well as} \quad \lim_{\Delta x \to 0} \frac{f(x+\Delta x) - f(x)}{\Delta x}$$

Although h might be "easier" to handle algebraically, we shall continue to use "delta notation" (Δx, Δt, etc.) in any situation where h would be ambiguous.

Key 21 Definition of derivative;

tangent to a graph

OVERVIEW *The process of finding an instantaneous rate of change function (see KEY 20-3) is so important in mathematics that it is given a name:* **differentiation**. *The resulting* **rate function** *is called the* **derivative**.

The derivative: To find the derivative of y with respect to x, denoted as $\dfrac{dy}{dx}$, evaluate the limit defined in KEY 21-1.

KEY DEFINITION 21-1

$$\frac{dy}{dx} = \lim_{h \to 0} \frac{f(x+h) - f(x)}{h}$$

- Compare this definition to KEY 20-1. Here, $h = \Delta x$ is unambiguous.

- The symbol $\dfrac{dy}{dx}$, due to Leibniz, is intended to be reminiscent of $\dfrac{\Delta y}{\Delta x}$.

- A word on notation: There are many symbols for the derivative of y with respect to x, each of which has advantages and disadvantages. Given $y = f(x)$, the symbols most often used are $\dfrac{dy}{dx}$, $\dfrac{df}{dx}$, $f'(x)$, y'_x, f'_x, $D_x y$, and $D_x f$. We shall freely use all of them. The x is commonly omitted from the last five when its existence is obvious (e.g., f').

- The derivative is the result of evaluating the limit of a special fraction (called a *difference quotient*) as $h \to 0$. If the limit exists at a particular value of x, we say that the function is **differentiable at x**; if the limit does not exist, the function is **nondifferentiable at x**. Therefore, differentiability, like continuity, is a *pointwise property*: it can exist at some values of x (points) and not at others.

- It is important to recall that, in order for any limit to exist, the left- and right-hand limits must exist and be equal. When differentiating, we may need to be sensitive to $h \to 0^+$ and $h \to 0^-$.

KEY EXAMPLE 21-2

Find the derivative $\left(\dfrac{df}{dx} \right)$ of the function $f(x) = x^2 - 6x + 5$.

Solution:

$$\frac{dy}{dx} = \lim_{h \to 0} \frac{f(x+h) - f(x)}{h} \qquad \text{(KEY 15-1)}$$

$$= \lim_{h \to 0} \frac{(x+h)^2 - 6(x+h) + 5 - (x^2 - 6x + 5)}{h}$$

$$= \lim_{h \to 0} \frac{\cancel{x^2} + 2hx + h^2 - \cancel{6x} - 6h + \cancel{5} - \cancel{x^2} + \cancel{6x} - \cancel{5}}{h}$$

$$= \lim_{h \to 0} \frac{\cancel{h}(2x + h - 6)}{\cancel{h}} = \lim_{h \to 0} (2x + h - 6) = 2x - 6$$

KEY EXAMPLE 21-3

If $y = \sqrt{x+2}$, find $\dfrac{dy}{dx}$.

Solution:

$$\frac{dy}{dx} = \lim_{h \to 0} \frac{y(x+h) - y(x)}{h} = \lim_{h \to 0} \frac{\sqrt{x+h+2} - \sqrt{x+2}}{h}$$

$$= \lim_{h \to 0} \frac{\sqrt{x+h+2} - \sqrt{x+2}}{h} \cdot \frac{\sqrt{x+h+2} + \sqrt{x+2}}{\sqrt{x+h+2} + \sqrt{x+2}}$$

(Rationalize the numerator.)

$$= \lim_{h \to 0} \frac{(x+h+2) - (x+2)}{h\left(\sqrt{x+h+2} + \sqrt{x+2}\right)} = \lim_{h \to 0} \frac{h}{h\left(\sqrt{x+h+2} + \sqrt{x+2}\right)}$$

$$= \lim_{h \to 0} \frac{1}{\sqrt{x+h+2} + \sqrt{x+2}} = \frac{1}{\sqrt{x+2} + \sqrt{x+2}} = \frac{1}{2\sqrt{x+2}}$$

KEY EXAMPLE 21-4

If $y = \dfrac{1}{x}$, find $\dfrac{dy}{dx}$.

Solution:

$$\frac{dy}{dx} = \lim_{h \to 0} \frac{f(x+h) - f(x)}{h} = \lim_{h \to 0} \frac{\dfrac{1}{x+h} - \dfrac{1}{x}}{h}$$

$$= \lim_{h \to 0} \frac{\dfrac{1}{x+h} - \dfrac{1}{x}}{h} \cdot \frac{x(x+h)}{x(x+h)} \quad \text{(Simplify the complex fraction.)}$$

$$= \lim_{h \to 0} \frac{x - (x+h)}{hx(x+h)} = \lim_{h \to 0} \frac{-\cancel{h}}{\cancel{h}x(x+h)} = \lim_{h \to 0} \frac{-1}{x(x+h)} = \frac{-1}{x^2}$$

- In the preceding examples, x is a constant variable and h is a true variable.
- To repeat a previous note on correct use of notation: The symbol "lim" is written until the process of taking the limit is performed, after which the symbol is omitted.

Geometric interpretation of the derivative: Consider the three diagrams shown in KEY 21-5.

KEY DIAGRAMS 21-5

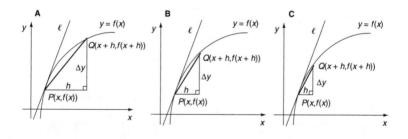

Note the following:

- Line ℓ is the tangent to the graph of $y = f(x)$ at point P.
- As $h \to 0$, point Q approaches point P along the graph and the slope of $\overline{PQ}$ approaches the slope of the tangent.
- By KEY 21-1, $\lim\limits_{h \to 0} m_{\overline{PQ}}$ = the value of the derivative at point P.

KEY NOTE 21-6

The value of the derivative at a point on the graph of a function is equal to the slope of the tangent to the graph at that point. The **slope of the tangent to the graph** at a point is also called the **slope of the curve (or graph)** at that point.

Before considering a classic problem in KEY 21-9, we discuss an alternative way of defining the derivative using different but frequently used notation.

An alternative (equivalent) definition of derivative: As we have already noted, the derivative of a function is a function. The value of the derivative at $x = a$, $f'(a)$, is given by KEY 21-7.

KEY DEFINITION 21-7

$$f'(a) = \lim_{x \to a} \frac{f(x) - f(a)}{x - a}$$

- Compare the two definitions of derivative (KEYS 21-1 and 21-7). A possible source of confusion may be that, in 21-1, x is a constant variable (h is the true variable) but in 21-7, x is the true variable (a is a constant variable).
- To transform 21-7 into 21-1: In 21-7, replace each x with $x + h$ and then each a with x. NOTE: $(x + h \to x) \Leftrightarrow (h \to 0)$.
- It may be useful to compare the diagram in KEY 21-8 with KEY 21-5A.

KEY DIAGRAM 21-8

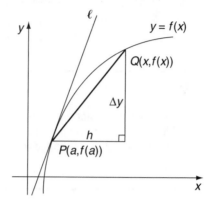

KEY EXAMPLE 21-9

Find the equation of the tangent to the graph of $y = f(x) = \sqrt{x+2}$ at point (7,3).

Solution: The slope of the tangent is given by the value of the derivative:

$$f'(x) = \frac{1}{2\sqrt{x+2}}$$

(See KEY 21-3.) Therefore, $f'(7) = \frac{1}{6}$.

Use the point-slope formula of a straight line, $y - y_1 = m(x - x_1)$, with $m = \frac{1}{6}$; then the equation of the tangent is

$$y - 3 = \frac{1}{6}(x - 7) \quad \text{or} \quad y = \frac{1}{6}x + \frac{11}{6}$$

KEY DEFINITION 21-10

A **normal** to a graph is the line perpendicular to the tangent to the graph at the point of tangency.

- Recall: The slopes of perpendicular lines are negative reciprocals.

KEY EXAMPLE 21-11

Find the equation of the normal to the graph of $f(x) = x^2$ at $x = 3$.

Solution: $f'(x) = 2x$. The slope of the tangent to the graph at $(3,9)$ is equal to $f'(3) = 6$. Therefore, the slope of the normal is $-\dfrac{1}{6}$. Use the point-slope formula of a straight line; then the equation of the normal is

$$y - 9 = -\frac{1}{6}(x - 3) \quad \Rightarrow \quad y = -\frac{1}{6}x + \frac{19}{2}$$

Key 22   Differentiation theorems

OVERVIEW *As we have seen, the process of differentiation can become complicated, even more so when the rule of the function is more complex. Fortunately, mathematicians have derived theorems (rules) that greatly simplify our work.*

KEY THEOREM 22-1

If a function is differentiable at $x = a$, then it is continuous at $x = a$.

- Whenever the derivative of a function exists, that function is guaranteed to be continuous. However, the converse of the theorem is not true. Consider the absolute-value function, which is continuous at $x = 0$ (KEY 18-12). We now attempt to evaluate the derivative of the absolute-value function at $x = 0$ by evaluating the left and right limits.

$$\frac{dy}{dx} = \lim_{h \to 0^+} \frac{f(0+h) - f(0)}{h}$$

$$= \lim_{h \to 0^+} \frac{|0+h| - |0|}{h}$$

$$= \lim_{h \to 0^+} \frac{|h|}{h}$$

$$= \lim_{h \to 0^+} \frac{h}{h} \quad (\text{since } h > 0)$$

$$= \lim_{h \to 0^+} 1$$

$$= 1$$

$$\frac{dy}{dx} = \lim_{h \to 0^-} \frac{f(0+h) - f(0)}{h}$$

$$= \lim_{h \to 0^-} \frac{|0+h| - |0|}{h}$$

$$= \lim_{h \to 0^-} \frac{|h|}{h}$$

$$= \lim_{h \to 0^-} \frac{h}{h} \quad (\text{since } h < 0)$$

$$= \lim_{h \to 0^+} (-1)$$

$$= -1$$

- Since the right-hand (limit) derivative is *not* equal to the left-hand (limit) derivative, the (limit) derivative does not exist at $x = 0$. Therefore, the absolute-value function at $x = 0$ is an example of a function that is continuous but not differentiable at a point.

- If a function is continuous at a point but the right-hand and left-hand derivatives are unequal, then the point is called a **cusp point**. A graph at a cusp point comes to a point, as opposed to being smooth, as shown in the accompanying diagram.

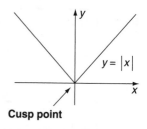

Cusp point

KEY THEOREM 22-2

If $f(x) = c$, where c is any constant real number, then $f'(x) = 0$.

- "The derivative of a constant function is equal to 0 at every real number." This makes sense because the slope of a constant function is always 0.

KEY EXAMPLE 22-3

If $f(x) = 7$, then $f'(x) = 0$.

KEY THEOREM 22-4 (power rule)

$D_x x^n = n x^{n-1}$ for all fixed values of n.

KEY EXAMPLES 22-5

By KEY 22-4

A. If $y = x^5$, then $y' = 5x^4$.

B. If $y = \sqrt{x} = x^{\frac{1}{2}}$, then $y' = \frac{1}{2} x^{-\frac{1}{2}} = \frac{1}{2} \cdot \frac{1}{x^{1/2}} = \frac{1}{2\sqrt{x}}$.

- Mathematics is the study of relationships, not names; the latter merely serves as a convenient way to refer to complex ideas. The derivative is dependent on the *relationship* between the variables, not their names. Therefore, $D_x x^5 = 5x^4$, $D_t t^5 = 5t^4$, and $D_f f^5 = 5f^4$. This is sometimes referred to as the "dummy variable" concept.

KEY THEOREMS 22-6 (sum/difference rules)

$$D_x(f(x)+g(x)) = f'(x)+g'(x)$$
$$D_x(f(x)-g(x)) = f'(x)-g'(x)$$

- "The derivative of a sum (difference) equals the sum (difference) of the derivatives."

KEY EXAMPLES 22-7

By KEY 22-6

A. $D_x(x^5 + x^3) = D_x x^5 + D_x x^3 = 5x^4 + 3x^2$
B. $D_x(x^5 - x^3) = D_x x^5 - D_x x^3 = 5x^4 - 3x^2$

KEY THEOREM 22-8 (product rule)

$$D_x[f(x) \cdot g(x)] = f(x) \cdot g'(x) + g(x) \cdot f'(x)$$

- "The derivative of a product of two functions equals the 'first' function times the derivative of the 'second' function plus the 'second' function times the derivative of the 'first' function."

KEY EXAMPLE 22-9

By KEY 22-8, if $y = x^3 \sqrt{x} = x^3 \cdot x^{\frac{1}{2}}$, then

$$y' = x^3 \cdot \frac{1}{2} x^{-\frac{1}{2}} + x^{\frac{1}{2}} \cdot 3x^2 = \frac{7}{2} x^{\frac{5}{2}}$$

- Yes, there is an easier way. Since $y = x^3 \cdot x^{\frac{1}{2}} = x^{\frac{7}{2}}$, by KEY 22-3,

$y' = \frac{7}{2} x^{\frac{5}{2}}$. However, as you will see later, an easier option is usually not available, so it is imperative that you be able to use the product rule.

KEY THEOREM 22-10 (generalized product rule)

$$D_x(f_1 f_2 f_3 \ldots f_n) = (f_1' f_2 f_3 \ldots f_n) + (f_2' f_1 f_3 \ldots f_n)$$
$$+ \ldots + (f_n' f_1 f_2 f_3 \ldots f_{n-1})$$

- "The derivative of a product of any number of functions is equal to a sum of terms *each of which* is the derivative of one of the factors multiplied by all of the remaining factors of the original product."

KEY EXAMPLE 22-11

(There will be more meaningful examples in later chapters.)

$$D_x(x^3 \cdot x^5 \cdot x^7) = 3x^2 \cdot x^5 \cdot x^7 + 5x^4 \cdot x^3 \cdot x^7 + 7x^6 \cdot x^3 \cdot x^5 = 15x^{14}$$

KEY THEOREM 22-12

$$D_x c f(x) = c D_x f(x)$$

- "The derivative of a constant times a function equals the constant times the derivative of the function."

KEY EXAMPLE 22-13

By KEY 22-12

$$D_x 6x^5 = 6 D_x x^5 = 6(5x^4) = 30x^4$$

KEY THEOREM 22-14 (quotient rule)

$$D_x \frac{f(x)}{g(x)} = \frac{g(x)f'(x) - f(x)g'(x)}{g^2(x)}$$

- "The derivative of a quotient of two functions is equal to the denominator times the derivative of the numerator minus the numerator times the derivative of the denominator, all divided by the square of the denominator."

KEY EXAMPLE 22-15

Find the derivative of $y = \dfrac{x^5}{x^2+3}$.

Solution: By KEY 22-14, if $y = \dfrac{x^5}{x^2+3}$, then

$$y' = \frac{(x^2+3)D_x x^5 - x^5 D_x(x^2+3)}{(x^2+3)^2}$$

$$= \frac{(x^2+3)5x^4 - x^5(2x+0)}{(x^2+3)^2} \quad \text{(the derivative)}$$

$$= \frac{5x^6 + 15x^4 - 2x^6}{(x^2+3)^2} \quad \text{(Now simplify the result.)}$$

$$= \frac{3x^6 + 15x^4}{(x^2+3)^2}$$

$$= \frac{3x^4(x^2+5)}{(x^2+3)^2} \quad \text{(This last step is frequently useful.)}$$

KEY EXAMPLE 22-16

At which points on the graph of $y = x^4 - 2x^2 + 1$ is the tangent parallel to the x-axis?

Solution: Since the slope of a line parallel to the x-axis is zero, find the derivative (which gives the slope) and set it equal to zero. Use the rules for differentiation; then

$$y' = 4x^3 - 4x \qquad \text{(the derivative)}$$

$$4x^3 - 4x = 0 \qquad \text{(Set the derivative equal to 0.)}$$

$$4x(x^2 - 1) = 4x(x+1)(x-1) = 0 \quad \text{(Solve by factoring.)}$$

The solution set is $\{0, 1, -1\}$, so the points are $(0,1)$, $(-1,0)$ and $(1,0)$.

KEY EXAMPLE 22-17

Find the x-coordinate of each point on the graph of $y = 1 - x^2$ at which the tangent line passes through point $(2,0)$.

Solution: You need to examine *all* of the tangents to the graph in order to determine which of them passes through the given point. At any point $(x_1, 1 - x_1{}^2)$ on the graph, the slope of the tangent is given by the value of the derivative at that point. Since $y' = -2x$, then $y'(x_1) = -2x_1$. Therefore, the equation of the tangent to the graph at point $(x_1, 1 - x_1{}^2)$, using the point-slope formula of a line, is

$$y - (1 - x_1^2) = -2x_1(x - x_1)$$

The question asks for which value(s) of x_1 will the tangent contain $(2,0)$. Therefore, substitute 2 for x and 0 for y and then solve for x_1:

$$0 - (1 - x_1^2) = -2x_1(2 - x_1) \qquad \text{(Substitute.)}$$

$$-1 + x_1^2 = -4x_1 + 2x_1^2$$

$$x_1^2 - 4x_1 + 1 = 0$$

$$x_1 = \frac{4 \pm \sqrt{12}}{2} = 2 \pm \sqrt{3} \quad \text{(quadratic formula)}$$

There are two points on the graph at which the tangents pass through point $(2,0)$, and their x-coordinates are $2 + \sqrt{3}$ and $2 - \sqrt{3}$.

- It is important to appreciate the difference between x and x_1. The variable x_1 represents the first coordinate of points on the parabola, and x represents the first (and y the second) coordinate of points on the tangent. The ability to understand and create such notation is invaluable in solving mathematics problems.

KEY EXAMPLE 22-18

Determine the value of the constant k such that the straight line determined by points $(0,3)$ and $(5,-2)$ is tangent to the graph of $y = \dfrac{k}{x+1}$.

Solution: Although the question asks only for the value of k, there is another unknown: the point of intersection. Usually, to find two unknowns one needs two equations. They, in turn, will be derived from two conditions. In this problem, the two conditions are as follows: at the point of tangency, (1) the y-coordinates are equal and (2) the slope of the graph (value of the derivative) equals the slope of the tangent.

$y = -x + 3$ is the equation of the line passing through $(5,-2)$ and $(0,3)$.

$y = \dfrac{k}{x+1} \Rightarrow y' = \dfrac{-k}{(x+1)^2}$ (quotient rule)

(or use the chain rule on $y = k(x+1)^{-1}$)

Let $x_1 =$ the first coordinate of the point of intersection; then:

$\dfrac{k}{x_1+1} = -x_1 + 3 \quad \Rightarrow \quad k = -x_1^2 + 2x_1 + 3$ (1) (y-coordinates equal)

$\dfrac{-k}{(x_1+1)^2} = -1 \quad \Rightarrow \quad k = x_1^2 + 2x_1 + 1$ (2) (slopes equal)

$x_1^2 + 2x_1 + 1 = -x_1^2 + 2x_1 + 3$

(First finding x_1 is easier than finding k.)

$2x_1^2 = 2 \Rightarrow x_1 = 1 \text{ or } x_1 = -1$

$\left(\text{Reject } -1; \text{ not in domain of } y = \dfrac{k}{x+1}.\right)$

Since $x_1 = 1$, return to one of the original equations to find $k = 4$.

Higher-order derivatives: Since the derivative of a function is itself a function, one can find the derivative of the derivative. The result is called the *second derivative* of the original function. There are also third derivatives, fourth derivatives, and, generally, nth derivatives.

NOTATION:

Second derivative:	Third derivative:
$\dfrac{d^2y}{dx^2}, f''(x), D_x^2 y, \text{etc.}$	$\dfrac{d^3y}{dx^3}, f'''(x), D_x^3 y, \text{etc.}$
Fourth derivative:	nth derivative:
$\dfrac{d^4y}{dx^4}, f^{(4)}(x), D_x^4 y, \text{etc.}$	$\dfrac{d^ny}{dx^n}, f^{(n)}(x), D_x^n y, \text{etc.}$

NOTE: Parentheses are required in $f^{(4)}(x)$ because $f^{(4)}(x)$ means "f raised to the fourth power."

KEY EXAMPLE 22-19

Find the third derivative of the function $f(x) = 2x^5 + x^4 - 2x^3 - 8x + 4$.

Solution: Use KEYS 22-6, 22-12, 22-4, and 22-2 (mentally in that order); then

$$f'(x) = 10x^4 + 4x^3 - 6x^2 - 8$$
$$f''(x) = 40x^3 + 12x^2 - 12x$$
$$f'''(x) = 120x^2 + 24x - 12$$

KEY EXAMPLE 22-20

Find the derivative at every real number for the "piecewise" function defined by

$$f(x) = \begin{cases} x^2 + 1 & \text{if} \quad x < 0 \\ -x^2 + 1 & \text{if} \quad 0 \le x \le 1 \\ 2x^2 - 2x & \text{if} \quad 1 < x \le 2 \\ x^3 & \text{if} \quad x > 2 \end{cases}$$

Solution: Use the rules for differentiating; then

$$f'(x) = \begin{cases} 2x & \text{if} & x < 0 \\ -2x & \text{if} & 0 < x < 1 \\ 4x - 2 & \text{if} & 1 < x < 2 \\ 3x^3 & \text{if} & x > 2 \end{cases}$$

NOTE: We have not yet specified the derivative at the "break points" ($x = 0$, $x = 1$, and $x = 2$), where the rule for the function changes, because continuity and the derivative must be examined carefully on either side of the break point.

- At $x = 0$: f is continuous (KEYS 14-2, 18-1, and 18-13); the **left derivative** is $2x$, and the **right derivative** is $(-2x)$. As $x \to 0$, the left and right derivatives each approach zero, so $f'(x)$ exists and $f'(0) = 0$.
- At $x = 1$: f is continuous because

$$\lim_{x \to 1^-} f(x) = \lim_{x \to 1^-} (-x^2 + 1) = 0,$$
$$\lim_{x \to 1^+} f(x) = \lim_{x \to 1^+} (2x^2 - 2) = 0, \text{ and } f(1) = 0$$

As $x \to 1$, the left derivative $(-2x)$ approaches **−1** and the right derivative $(-4x)$ approaches **−4**, so the derivative fails to exist. However, since f is **continuous at $x = 1$**, there is a **cusp point** at $x = 1$.

- At $x = 2$: As $x \to 2$, the left derivative ($4x - 2$) approaches **6** and the right derivative ($3x^2$) approaches **12**, so the derivative fails to exist. However, there is no continuity and therefore no cusp point since

$$\lim_{x \to 2^-} (2x^2 - 2) = 6 \neq \lim_{x \to 2^+} x^3 = 8$$

NOTE: By KEY 22-1, discontinuity itself implies lack of differentiability.

Key 23 Chain rule; differentiation of implicit functions

OVERVIEW *In Key 22 we listed rules for differentiation that are applicable to certain "simple" situations. But what if the function to be differentiated is a **composite** of these situations? Also, is it possible to find the derivative of an **implicit function** (one that cannot be solved explicitly for one of the variables)? The **chain rule** to the rescue!*

Important observation: Formulas are derived from *relationships*, not the names of objects. For example, the power rule, $D_x x^n = n x^{n-1}$, does not say "The derivative of x to the nth power with respect to x equals n times x to the $n - 1$ power." What it does say is "The derivative (instantaneous rate of change) of *any* variable raised to the nth power with respect to *that* variable is n times *that* variable to the $n - 1$ power." Hence

$$D_x x^5 = 5x^4, \quad D_f f^5 = 5f^4, \quad \text{and} \quad D_{x^2+1}(x^2+1)^5 = 5(x^2+1)^4$$

However, the power rule does not apply to $D_x(x^2 + 1)^5$ because the relationship between $(x^2 + 1)^5$ and x is not a simple power relationship, as indicated by the examples above. For $D_x(x^2 + 1)^5$, we need the **chain rule**.

KEY THEOREM 23-1 (chain rule)

$$D_x g(f(x)) = D_{f(x)} g(f(x)) \cdot D_x f(x)$$

- In a composite of two functions, that is, $g(f(x))$, $f(x)$ is commonly called the *inner function* and $g(x)$ the *outer function*. The chain rule, in words, says "The derivative of a function with respect to a variable with which it has a composite relationship equals the derivative of the outer function (the entire composite) with respect to the inner function times the derivative of the inner function with respect to the original variable."
- Finding the derivative of one function with respect to another function sounds more complicated than it is. For example, the

derivative of $(x^2 + 1)^5$ with respect to $(x^2 + 1)$ is $5(x^2 + 1)^4$ because the function $(x^2 + 1)^5$ and the function $(x^2 + 1)$ have a (fifth) *power* relationship, making the power rule applicable.

- Here is another way of writing the chain rule: $\dfrac{dg}{dx} = \dfrac{dg}{df}\dfrac{df}{dx}$.

  Notice that these symbols for the derivatives seem to act like fractions. With appropriate definitions, they really will!

KEY EXAMPLE 23-2

If $y = (x^2 + 1)^5$, find $\dfrac{dy}{dx}$.

Solution: By KEY 23-1

$$\frac{dy}{dx} = D_{x^2+1}\left(x^2+1\right)^5 \cdot D_x\left(x^2+1\right)$$

$$= 5\left(x^2+1\right)^4(2x) = 10x(x^2+1)^4$$

KEY EXAMPLE 23-3

If $y = \sqrt{x^2+1}$, find $\dfrac{dy}{dx}$.

Solution: By KEY 23-1

$$\frac{dy}{dx} = D_{x^2+1}\left(x^2+1\right)^{\frac{1}{2}} \cdot D_x\left(x^2+1\right)$$

$$= \frac{1}{2}\left(x^2+1\right)^{-\frac{1}{2}}(2x) = x(x^2+1)^{-\frac{1}{2}}$$

- Notice the structural similarities of KEYS 22-2 and 22-3. We can generalize and write a special case of the chain rule.

KEY THEOREM 23-4 (generalized power rule)

$$D_x f^n(x) = nf^{n-1}(x) \cdot D_x f(x)$$

Derivatives of implicit functions: To find the derivative of a function not explicitly solved for one of its variables, differentiate *both sides of the equation* with respect to the appropriate variable, using applicable rules.

KEY EXAMPLE 23-5

Find the equation of the tangent to the graph of $x^3 - xy + y^3 = 7$ at $(1,2)$ on the graph.

Solution:

$$D_x(x^3 - xy + y^3) = D_x 7$$

$$D_x x^3 - D_x xy + D_x y^3 = 0 \qquad \text{(KEYS 22-6 and 22-2)}$$

$$D_x x^3 - (xD_x y + yD_x x) + D_y y^3 \cdot D_x y = 0 \qquad \text{(KEYS 22-8 and 23-1)}$$

$$3x^2 - \left(x\frac{dy}{dx} + y(1)\right) + 3y^2 \frac{dy}{dx} = 0 \qquad \text{(KEY 22-4)}$$

$$3x^2 - x\frac{dy}{dx} - y + 3y^2 \frac{dy}{dx} = 0 \qquad \left(\text{Solve for } \frac{dy}{dx}.\right)$$

$$-x\frac{dy}{dx} + 3y^2 \frac{dy}{dx} = y - 3x^2$$

$$\frac{dy}{dx}(-x + 3y^2) = y - 3x^2$$

$$\frac{dy}{dx} = \frac{y - 3x^2}{-x + 3y^2} \quad \text{(the derivative)}$$

$$\left.\frac{dy}{dx}\right|_{(1,2)} = \left.\frac{y - 3x^2}{-x + 3y^2}\right|_{(1,2)} = \frac{2 - 3(1^2)}{-1 + 3(2^2)}$$

$$= -\frac{1}{11} \quad \text{(the slope of the tangent)}$$

Therefore, the equation of the tangent is $y - 2 = -\dfrac{1}{11}(x - 1)$.

- $\left.\dfrac{dy}{dx}\right|_{(1,2)}$ means "$\dfrac{dy}{dx}$ evaluated at point $(1,2)$."
- When we differentiate an implicit function (as opposed to an explicit function), the *dependent* variable (in this case, y) appears in the rule for the derivative. In many cases, both x and y appear in the derivative.

Key 24 Linear approximations using differentials

OVERVIEW *What if we wish to evaluate a function at a number at which the calculations are difficult? It is possible to approximate the function using a line. However, the best results are obtained when the line is tangent to the graph of the function and the point of approximation is "close" to the point of tangency.*

KEY DEFINITION 24-1

When calculating the change in a *function* y, for a particular change in x, we use the symbols Δy and Δx. When calculating the change in y along the *tangent* to the graph of the function for the same change in x, we use the symbols dy and dx, which are called **differentials**.

KEY DIAGRAM 24-2

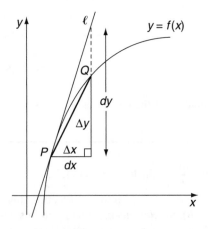

- The use of straight lines to approximate values of functions has been made obsolete by sophisticated calculators that easily determine these values with excellent accuracy. However, linear approximations play an important role in mathematics and science. KEY 24-3 clarifies the situation.
- From KEY 24-1, $m_\ell = dy \div dx$; and from KEY 21-5, $m_\ell = f'(x)$. Therefore, $dy \div dx = f'(x)$ and that implies $dy = f'(x) \cdot dx$.
- The preceding observation about the differentials dx and dy helps to clarify certain symbolic manipulations that we shall be doing in future topics.
- If $y = x^3$, then $dy = 3x^2\, dx$.

KEY EXAMPLE 24-3

Use differentials to approximate $\sqrt[4]{15.95}$.

Solution: $\sqrt[4]{16} = 2$. Using $\Delta x = -0.05$, you wish to calculate Δy, the change in y along the graph of the function $y = \sqrt[4]{x} = x^{\frac{1}{4}}$. Therefore, approximate Δy with dy, the change in y along the tangent to the graph, at $x = 16$, the "anchor point." Then

$$dy = f'(x)\, dx \Rightarrow dy = \frac{1}{4} x^{-\frac{3}{4}}\, dx$$

$$dy\big|_{\substack{x=16 \\ dx=-0.05}} = \frac{1}{4}\left(16^{-\frac{3}{4}}\right)(-0.05)$$

$$= \frac{1}{4}\left(\frac{1}{8}\right)(-0.05) = -0.0015625$$

Therefore, $\sqrt[4]{15.95} \approx 2 + (-0.0015625) = 1.9984375$.

- A calculator gives $\sqrt[4]{15.95} \approx 1.998435666$.
- The size of the error in the approximation depends on the shape of the graph (how closely the tangent approximates the graph of the function) and the size of dx (how close the point of approximation is to the point of tangency—the "anchor point").

Theme 5 APPLICATIONS OF THE DERIVATIVE

*I*n Theme 4, we observed that the derivative is the instantaneous rate of change of a function, which is geometrically given by the slope of the tangent to its graph. This information can be used to determine the shape of the graph, including the locations of "peaks" and "valleys." Therefore, we can use the derivative to determine the **maximum and minimum values of a function**. In addition, rates of change may be used to describe the motion of an object.

Key 25 Related rates

OVERVIEW *When information about different variables is given in terms of their derivatives, the analysis is called a **related-rates problem**. The relationship between (among) the variables themselves is usually implicit in the problem.*

How to solve a related-rates problem:

1. If relevant, draw a diagram depicting the state of affairs at a general time (not at any one specific time) and name the variables.
2. List all given rates and unknown rates, using derivatives.
3. Determine the relationship, an equation, between (among) the variables from the conditions of the problem.
4. Resist the temptation to substitute given values for the variables until after step 5 (otherwise, you will "lose" the function defining the relationship).
5. Differentiate both sides of the equation obtained in step 3 with respect to the common independent variable in the problem, usually time, t.
6. Substitute the given information into the resulting equation, and determine the unknown.

KEY OBSERVATION 25-1

If variable u is a function of variable v and v is a function of variable t, then $\dfrac{du}{dt} = \dfrac{du}{dv}\dfrac{dv}{dt}$ (the chain rule).

- If I bake $3\dfrac{\text{pies}}{\text{hr}}$ and sell them at $9\dfrac{\$}{\text{pie}}$, then I make $3\dfrac{\text{pies}}{\text{hr}} \cdot 9\dfrac{\$}{\text{pie}} = 27\dfrac{\$}{\text{hr}}$.

KEY EXAMPLE 25-2

If $u = v^3$, then $\dfrac{du}{dt} = 3v^2\dfrac{dv}{dt}$.

KEY EXAMPLE 25-3

Air is being pumped into a spherical balloon at the rate of 100 cubic centimeters per second. How fast is the radius of the balloon increasing when the radius is 5 centimeters?

Solution: The volume of the balloon is increasing at the constant rate of $\dfrac{dV}{dt} = 100 \text{ cm}^3/\text{s}$. You need to evaluate $\dfrac{dr}{dt}\Big|_{r=5}$. In this problem, the relationship between the variables is a known formula for the volume of a sphere.

$$V = \frac{4}{3}\pi r^3 \qquad \text{(formula for the volume of a sphere)}$$

$$\frac{dV}{dt} = \frac{dV}{dr}\frac{dr}{dt} \qquad \text{(chain rule; differentiating both sides with respect to } t\text{)}$$

$$\frac{dV}{dt} = 4\pi r^2 \frac{dr}{dt} \qquad (V \text{ and } r \text{ are } implicit \text{ functions of time.)}$$

$$100 = 4\pi(5^2)\frac{dr}{dt}\Big|_{r=5} \qquad \text{(Substitute.)}$$

$$\frac{dr}{dt}\Big|_{r=5} = \frac{1}{\pi} \approx 0.318 \text{ cm}/\text{s} \quad \text{(Solve for } dr/dt.\text{)}$$

KEY EXAMPLE 25-4

A girl is flying a kite that is moving horizontally at the rate of 3 feet per second away from the girl at an altitude of 50 feet. How fast is the string unwinding when the kite is 130 feet from the girl?

Solution: Draw a diagram.

Given $\dfrac{dx}{dt} = 3 \text{ ft}/\text{sec}$, find $\dfrac{ds}{dt}\Big|_{s=130}$.

$$x^2 + 50^2 = s^2 \qquad \text{(Pythagorean Theorem)}$$

$$\cancel{2}x\frac{dx}{dt} = \cancel{2}s\frac{ds}{dt} \quad \text{(chain rule)}$$

$$(120)(3) = (130)\frac{ds}{dt}\Big|_{s=130} \qquad (s = 130 \Rightarrow x^2 + 50^2 = 130^2 \Rightarrow x = 120)$$

$$\frac{ds}{dt}\Big|_{s=130} = \frac{36}{13} \approx 2.769 \text{ ft/sec}$$

KEY EXAMPLE 25-5

A full conical reservoir (point down) is 80 feet deep and 240 feet wide at the top. As a result of evaporation, the depth of the water is decreasing at the rate 0.01 feet per day. What is the rate of evaporation when the reservoir is half full?

Solution: Draw a diagram

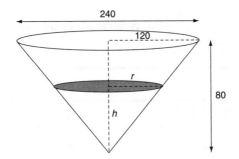

Given $\dfrac{dh}{dt} = -0.01$ ft/day, find $\dfrac{dV}{dt}\Big|_{h=40}$.

$$\left. \begin{array}{l} V = \dfrac{1}{3}\pi r^2 h \quad \text{(volume of a cone)} \\[2mm] \dfrac{r}{h} = \dfrac{120}{80} = \dfrac{3}{2} \quad \text{(similar triangles)} \end{array} \right\} \Rightarrow V = \dfrac{3}{4}\pi h^3$$

$$\frac{dV}{dt} = \frac{9}{4}\pi h^2 \frac{dh}{dt} \quad \text{(chain rule)}$$

$$\frac{dV}{dt}\Big|_{h=40} = \frac{9}{4}\pi(40^2)(-0.01) = -36\pi \approx -113.097 \text{ ft}^3/\text{day}$$

- Here, $\dfrac{dh}{dt}$ is negative because the depth of the water is *decreasing*.

Key 26 Graphing functions

OVERVIEW *If a rate of change is positive, then the dependent variable and its graph are* **increasing**; *if a rate of change is negative, then the dependent variable and its graph are* **decreasing**. *Also, whether the rate of change (the derivative) is increasing or decreasing affects how the graph is bending (its* **concavity***).*

KEY DEFINITIONS 26-1

A. A function $y = f(x)$ is **increasing on an interval** if and only if $x_2 > x_1 \Rightarrow f(x_2) > f(x_1)$.

B. A function $y = f(x)$ is **decreasing on an interval** if and only if $x_2 > x_1 \Rightarrow f(x_2) < f(x_1)$.

KEY THEOREMS 26-2

A. If, for all x in an interval, $f'(x) > 0$, then f is increasing on the interval.

B. If, for all x in an interval, $f'(x) < 0$, then f is decreasing on the interval.

- The converses of these theorems are not true. For example, f increasing does not guarantee that $f' > 0$. Consider the graph of $f(x) = x^3$ at the origin. The function is increasing in its entire domain (this should be clear) and, therefore, is increasing on an interval containing $x = 0$. But $f'(x) = 3x^2$ and $f'(0) = 0$ (not $f'(0) > 0$).

KEY DIAGRAM 26-3

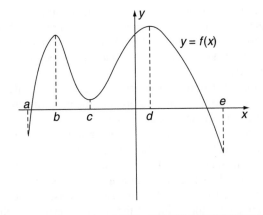

- The function f is increasing on the intervals (a, b) and (c, d). It is decreasing on the intervals (b, c) and (d, e).
- Recall that the value of the derivative is equal to the slope of the tangent to the graph (KEY 21-6). In the graph, observe that the function is increasing if and only if its derivative (slope) is positive and is decreasing if and only if its derivative (slope) is negative.

KEY DIAGRAM 26-4

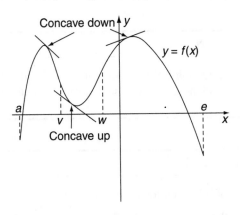

- Observe that, at every point in the interval (v, w), the tangent to the graph is below the graph on either side of the point of tangency; whereas, at every point in the intervals (a, v) and (w, e), the tangent to the graph is above the graph on either side of the point of tangency. This is caused by the way the graph is "bending."
- **Concavity:** In the interval (v, w) the graph is said to be **concave upward**. In the intervals (a, v) and (w, e) the graph is **concave downward**.
- Note that concavity is completely independent of increasing and decreasing. For example, a graph may be increasing or decreasing or both while being concave upward.
- Furthermore, observe that on the interval where the graph is concave upward, the slope of the tangent (not necessarily the function itself) is increasing. On the intervals where the graph is concave downward, the slope of the tangent is decreasing.
- Whether the slope (the derivative) is increasing or decreasing is determined by the value of the slope's derivative, that is, the derivative of the derivative, the second derivative of the function (see paragraph following KEY 22-18). This suggests the theorems in KEY 26-5.

KEY THEOREMS 26-5

A. If, for all x in an interval, $f''(x) > 0$, then f is concave upward on the interval.

B. If, for all x in an interval, $f''(x) < 0$, then f is concave downward on the interval.

KEY DEFINITION 26-6

The function f has a **point of inflection** at $x = k$ if and only if f is continuous at $x = k$ and the graph switches concavity around that point.

KEY EXAMPLE 26-7

Find all points of inflection for $y = x^3 - x^2$.

Solution: Find the second derivative:

$$y = x^3 - x^2 \Rightarrow y' = 3x^2 - 2x \Rightarrow y'' = 6x - 2$$

Observe that $y'' > 0$ for all $x > \dfrac{1}{3}$, $y'' < 0$ for all $x < \dfrac{1}{3}$, and the function y is continuous at $x = \dfrac{1}{3}$. (Polynomial functions are continuous at every real number.) Therefore, there is only one point of inflection, at $x = \dfrac{1}{3}$.

KEY METHODS 26-8

A. **How to determine intervals of increasing and decreasing:** Determine where f' is positive and where f' is negative.
B. **How to determine concavity:** Determine where f'' is positive and where f'' is negative.
C. **How to locate points of inflection:** First determine all values of x at which the graph is continuous and the second derivative is either 0 or nonexistent. Then, at each value, determine whether or not the second derivative changes sign.

KEY EXAMPLE 26-9

(a) Determine the intervals on which the function $f(x) = x^5 - 30x^3$ is increasing, decreasing, concave upward, and concave downward.
(b) Determine the points of inflection.

Solutions: (a) Start by finding the first and second derivatives and their zeros, factoring completely, if possible.

$$f'(x) = 5x^4 - 90x^2 = 5x^2(x^2 - 18) = 5x^2(x - \sqrt{18})(x + \sqrt{18})$$
$$f''(x) = 20x^3 - 180x = 20x(x^2 - 9) = 20x(x - 3)(x + 3)$$
$$f'(x) = 0 \iff x = 0 \text{ or } x = \pm\sqrt{18}$$
$$f''(x) = 0 \iff x = 0 \text{ or } x = \pm3$$

Now analyze where the functions are positive and where negative, using a number line (KEYS 1-3 and 1-4).

$$f'(x) = 5x^2(x - \sqrt{18})(x + \sqrt{18})$$

$$f''(x) = 20x(x - 3)(x + 3)$$

From f', the function f is increasing on $(-\infty, -\sqrt{18}) \cup (\sqrt{18}, \infty)$ and is decreasing on $(-\sqrt{18}, \sqrt{18})$. From f'', the function's graph is concave upward on $(-3, 0) \cup (3, \infty)$ and concave downward on $(-\infty, -3) \cup (0, 3)$. (b) There are points of inflection at points $(-3, 324)$, $(0, 0)$, and $(3, -324)$. Note the continuity and the change in sign of the second derivative at each of these points.

* Although $f'(0) = 0$ (not $f'(0) < 0$), 0 *is* included in the interval of decreasing behavior. Intuitively, if there is continuity and the function is decreasing both left and right of the point, then it is decreasing *at* the point (the function cannot stop decreasing for just one point). (See KEY 26-2.)

KEY EXAMPLE 26-10

Sketch the graph of the function $y = x^3 - 3x^2$.

Solution: Find the first and second derivatives.

$$y = x^3 - 3x^2$$
$$y' = 3x^2 - 6x = 3x(x - 2) \qquad y' = 0 \iff x = 0 \text{ or } x = 2$$
$$y'' = 6x - 6 \qquad y'' = 0 \iff x = 1$$

Next, use number lines to indicate the sign of each derivative.

$$y' = 3x(x - 2)$$

$$y'' = 6x - 6$$

From the number lines, you can see the intervals on which the graph is increasing, decreasing, concave upward, and concave downward.

Important points on the graph are (0,0), (2,–4), and (1,–2), the last point being a point of inflection.

The completed graph of the function $y = x^3 - 3x^2$ is shown below.

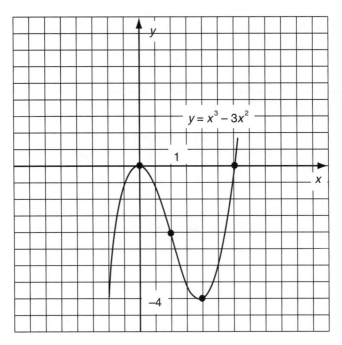

KEY EXAMPLE 26-11

Find all points of inflection for the graph of $y = \dfrac{1}{x}$.

Solution: Find the second derivative.

$$y = \frac{1}{x} = x^{-1} \quad \Rightarrow \quad y' = -x^{-2} = -\frac{1}{x^2} \quad \Rightarrow \quad y'' = 2x^{-3} = \frac{2}{x^3}$$

Observe that $y'' > 0 \Leftrightarrow x > 0$ and $y'' < 0 \Leftrightarrow x < 0$. However, there is no point of inflection at $x = 0$; the graph is discontinuous at $x = 0$ (in fact, 0 is not in its domain). Therefore, the graph of $y = \dfrac{1}{x}$ does not have a point of inflection.

Key 27 Relative maxima and minima

OVERVIEW *Using the discussion of KEY 26, we now are able to define "peaks and valleys" of a graph and determine their locations.*

KEY DEFINITIONS 27-1

A. A function f has a **relative maximum** at $x = a$ if and only if there exists an open interval containing a such that, for all x in the interval,

$$f(x) \leq f(a)$$

B. A function f has a **relative minimum** at $x = a$ if and only if there exists an open interval containing a such that, for all x in the interval,

$$f(x) \geq f(a)$$

C. A **relative extremum** is either a relative maximum or a relative minimum.

- If a function is never constant, then a relative maximum is a "peak" and a relative minimum is a "valley" in its graph.
- WARNING: Be sensitive as to whether the discussion is concerned with *where* the relative extremum is (at $x = a$) or *what* the relative extremum is ($f(a)$, the value of the function).

KEY DIAGRAM 27-2

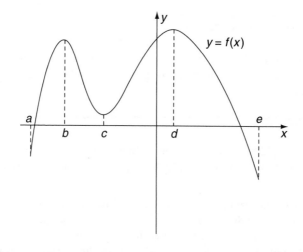

- The graph of $y = f(x)$ has relative maxima at $x = b$ and $x = d$. It has a relative minimum at $x = c$.
- A relative minimum is a "local" minimum. In the graph of $y = f(x)$, the function has smaller values than $f(c)$ elsewhere in the domain (e.g., at $x = e$). Note, however, that one of the relative (local) maxima also appears to be the largest value of the function in its domain, $[a, e]$.
- It appears that the slope of the tangents to the graph at the relative extrema is 0. Is this true for all functions? No. Consider $y = |x|$. There is a relative minimum at $x = 0$, but the derivative is undefined there (a cusp point; see the paragraphs following KEY 22-1), so there is no tangent.

KEY THEOREM 27-3

If a function f has a relative extremum at $x = a$, then either $f'(a) = 0$ or $f'(a)$ does not exist.

- Be careful; the converse is not true. Consider the graph of $y = x^3$ at the origin. Recall that, in the paragraph following KEY 26-2, we demonstrated that $y = x^3$ is an increasing function throughout its domain (all reals) and, therefore, cannot have any relative extremum, particularly not at $x = 0$. However, $y'(0) = 0$.

KEY DEFINITION 27-4

A **critical number** of a function f is any number a in the domain of the function such that the derivative at a is either equal to 0 or is nonexistent.

- It should be clear that relative extrema must occur at critical numbers, but not all critical numbers produce relative extrema. (See KEY 27-3.)

How to locate relative maxima and minima: First list all critical numbers by finding all numbers in the domain that make the derivative 0 or nonexistent. Then, to each critical number, apply the first-derivative test or the second-derivative test (your choice), as described below.

KEY THEOREM 27-5 (first-derivative test for relative extrema)

Suppose that c is a critical number of a function f and that the interval $[a, b]$, a subset of the domain, contains c and no other critical numbers. Then:

1. If $f'(a) > 0$ and $f'(b) < 0$, f has a relative maximum at $x = c$.
2. If $f'(a) < 0$ and $f'(b) > 0$, f has a relative minimum at $x = c$.
3. Otherwise, there is no relative extremum at $x = c$.

- Basically, if c is a critical number and the function is increasing to the left of c and decreasing to the right of c, then there must be a relative maximum at $x = c$. If the function is decreasing to the left of c and increasing to the right of c, then there must be a relative minimum at $x = c$.

KEY THEOREM 27-6 (second-derivative test for relative extrema)

Suppose that c is a critical number of a function f. Then:

1. If $f''(c) < 0$, f has a relative maximum at $x = c$.
2. If $f''(c) > 0$, f has a relative minimum at $x = c$.
3. If $f''(0) = 0$, the test fails (no conclusion can be drawn).

- To better understand the test, look at KEY 27-2; observe that relative maxima seem to occur where the graph is concave downward and relative minima to occur where the graph is concave upward.
- The failure of the test in the third case is demonstrated by example. Consider $f(x) = x^4$, $g(x) = -x^4$, and $h(x) = x^3$. For each, the second derivative is equal to 0 at the origin (at $x = 0$). Yet, at the origin, f is concave upward, g is concave downward, and h has a point of inflection (no concavity). For example, $f''(x) = 12x^2$, which is positive for all $x \neq 0$. Therefore, f is concave upward everywhere and cannot stop its concavity at a single point (the origin) since f is continuous. Clearly, no conclusion can be drawn when the second derivative is equal to 0 at a critical number.
- Note that you may "waste" some time if the second-derivative test fails; the first-derivative test never fails.
- **How to choose:** One test may be more difficult to apply than the other. The first-derivative test essentially requires a number-line analysis. If the second derivative is easy to find, then that test is easily applied; but, as stated above, it may fail. Your ability to choose the easier method will improve with experience.

KEY EXAMPLE 27-7

Find the relative extrema of $y = x^5 - 15x^3$.

Solution: $y' = 5x^4 - 45x^2$. The derivative is never nonexistent, so find all critical numbers by solving $5x^4 - 45x^2 = 5x^2(x^2 - 9) = 0$. The critical numbers are -3, 0, and 3. Now choose between two methods.
METHOD 1: Evaluate the second derivative at each critical number, and use the second-derivative test to determine the type of relative extremum, if any, in each case. Since $y'' = 20x^3 - 90x$:

$y''(-3) < 0 \implies$ a relative maximum at $x = -3$

$y''(0) = 0 \implies$ The text fails; you must use the first-derivative test.

$y''(3) > 0 \implies$ a relative minimum at $x = 3$

METHOD 2: Draw a number line to determine in which intervals the derivative is positive and in which it is negative.

$y'(x) = 5x^2(x - 3)(x + 3)$

Use the first-derivative test:

1. Because $y'(x) > 0$ to the left of -3 and $y'(x) < 0$ to the right of -3, there is a relative maximum at $x = -3$.
2. Because $y'(x) < 0$ to the left and to the right of 0, there is no relative extremum at $x = 0$.
3. Because $y'(x) < 0$ to the left of 3 and $y'(x) > 0$ to the right of 3, there is a relative minimum at $x = 3$.

- Notice that the two tests yielded significantly different results at $x = 0$. The second-derivative test *failed* to give a result. The first-derivative test gave a *definitive* result: no relative extremum.
- This caution bears repeating: Be careful with details, and do not reason by the converse. For example, although every relative extremum must occur at a critical number, not every critical number produces a relative extremum.

Key 28 Endpoint extrema and absolute extrema

OVERVIEW *In KEY 27, we discussed how to locate relative (local) extrema (if they exist). We are now ready to find the largest and smallest values (if they exist) of a function in its entire domain. We start by examining the behavior of a function at any endpoint of its domain.*

KEY DEFINITION 28-1

The value of a function at every endpoint (if there are any) of the domain is an **endpoint extremum**.

- WARNING: The term *endpoint extremum* refers to the *value of the function* (not to the value of x at the endpoint).

KEY DEFINITIONS 28-2

A. If there exists a number a in any domain of f such that, for all x in the domain, $f(a) \geq f(x)$, then $f(a)$ is the **absolute maximum** of f in its domain.
B. If there exists a number a in any domain of f such that, for all x in the domain, $f(a) \leq f(x)$, then $f(a)$ is the **absolute minimum** of f in its domain.
C. An **absolute extremum** is an absolute maximum or an absolute minimum.

- A function may achieve its absolute maximum (or minimum) at more than one location. In addition, there is no guarantee that a function will have an absolute maximum or minimum in a specified domain. (Nor is there a guarantee for relative extrema. Consider $y = x$ on the interval [1, 2].)

KEY THEOREM 28-3 (Extreme-Value Theorem)

If a function is continuous on a closed interval, then it must have an absolute maximum and an absolute minimum achieved in the interval.

- It is important to understand why the interval has to be closed in order to guarantee absolute extrema. First, try to sketch a graph that is continuous on a *closed* interval and fails to have both absolute extrema. Now consider the function $y = \dfrac{1}{x}$ on the *open* interval (0, 1). Although the function is continuous on the interval, it fails to have an absolute maximum (the function grows infinitely large) and absolute minimum (try naming the smallest value). Of course, a function *may* have absolute extrema on an open interval.

KEY THEOREM 28-4 (Intermediate-Value Theorem)

If a function f is continuous on a closed interval $[a, b]$ and k is any number between $f(a)$ and $f(b)$, then there exists at least one number c between a and b such that $f(c) = k$.

- The Intermediate-Value Theorem guarantees that a continuous function achieves every value between its absolute maximum and absolute minimum somewhere in its domain.
- A useful consequence: If f is continuous on $[a, b]$ and if $f(a)$ and $f(b)$ have opposite signs, then there is at least one solution to the equation $f(x) = 0$ in the interval (a, b). For example, because $f(x) = x^2 - 2$ is continuous on $[1, 2]$, $f(1) < 0$, and $f(2) > 0$, the equation $x^2 - 2 = 0$ has at least one solution in the interval $(1,2)$. (Of course it is $\sqrt{2} \approx 1.414$.)

KEY THEOREMS 28-5

If a continuous function f has exactly one critical number a, then

A. If $f(a)$ is a relative maximum, $f(a)$ is an absolute maximum.
B. If $f(a)$ is a relative minimum, $f(a)$ is an absolute minimum.

- KEY 28-5 should be intuitively clear. If a continuous function has exactly one critical number and it produces a relative maximum, for example, then that function can have no relative minima (no other critical numbers). Without relative minima, the graph cannot "turn around" to eventually have larger images than the relative maximum.

KEY METHOD 28-6

How to locate and evaluate absolute extrema (if they exist) of a function

1. Find all critical numbers (KEY 27-4).
2. For each critical number use the first-derivative test or the second-derivative test to determine whether, at that number, the function has a relative maximum, a relative minimum, or neither (KEYS 27-5 and 27-6).
3. Evaluate the function at each endpoint of its domain (endpoint extrema).
4. To find the absolute maximum, compare the values of the function's relative maxima and endpoint extrema and choose the largest value. To find the absolute minimum, compare the values of the function's relative minima and endpoint extrema and choose the smallest value.

- **See KEY 28-9 for an important shortcut method.**

KEY EXAMPLE 28-7

Find the absolute extrema of the function

$$y = x^2 - 6x + 3.$$

Solution: First examine the first derivative to find critical numbers.
$y = x^2 - 6x + 3 \Rightarrow y' = 2x - 6$ (Note that the derivative is never nonexistent.)

$2x - 6 = 0 \Rightarrow x = 3$. Therefore, 3 is the only critical number.

$y'' = 2 \Rightarrow y''(3) = 2 > 0$. By the second-derivative test, there is a relative minimum at $x = 3$.

Since the function is continuous (its derivative exists at all real numbers), you can assert that there is an *absolute minimum* at $x = 3$ (KEY

28-5B). Moreover, the domain is the set of all real numbers, so there are no endpoints. The values of this function grow infinitely large as x grows infinitely large (the graph of the function is a parabola with a minimum vertex). Therefore, this function has no absolute maximum.

KEY EXAMPLE 28-8

Find the absolute extrema of the function $y = x^3 - 3x^2$ over the interval $[-1, 3.1]$.

Solution: First observe that the function is polynomial and therefore continuous over the closed interval. Therefore, it must have an absolute maximum and an absolute minimum. Examine the first derivative to find critical numbers of the function.

$y = x^3 - 3x^2 \Rightarrow y' = 3x^2 - 6x = 3x(x - 2)$

$3x(x - 2) = 0 \Rightarrow x = 0$ or $x = 2$. There are two critical numbers, 0 and 2.

$y'' = 6x - 6$. Use the second-derivative test for relative extrema; then

$y''(0) = -6 < 0 \Rightarrow$ There is a relative maximum at $x = 0$.

$y''(2) = 6 > 0 \Rightarrow$ There is a relative minimum at $x = 2$.

Now evaluate the function's relative extrema and endpoint extrema.

x	$f(x)$	Comment
−1	−4	endpoint extremum
0	0	relative maximum
2	−4	relative minimum
3.1	0.961	endpoint extremum

By observation, the absolute maximum is 0.961 (at $x = 3.1$) and the absolute minimum is −4 (twice: at $x = -1$, an endpoint, and at $x = 2$, a critical number).

- The graph of this function is found in KEY 26-10. You should check to see whether all of the conclusions drawn above are consistent with the graph. Also note that you could have used the first-derivative test (number line in KEY 26-10) to determine relative extrema.

KEY METHOD 28-9 (a shortcut)

If the only objective is to find absolute extrema and we are certain they exist, there is a quicker, if less refined, procedure: Do not bother with the first-derivative or second-derivative test for relative extrema. Just generate a table of values, including all critical numbers and endpoints. Since absolute extrema occur only at critical numbers and endpoints, we can get all required information from the table.

Key 29 Maximizing and minimizing functions

OVERVIEW *Maximizing and minimizing functions is more than pure mathematics; there are many practical applications. Why are all 6-ounce tuna-fish cans the same shape and size (congruent)? The reason involves more than product identification. The dimensions of the can are chosen so that the surface area (material required and cost to manufacture) is minimized. In this section, we solve some classic "max-min" problems.*

KEY EXAMPLE 29-1 (a simple rectangle problem)

A stone wall will be used for one side of a rectangular plot. What is the maximum area that can be enclosed with 400 feet of fencing?

Solution: A diagram will help. Let w = the width of the plot, and l = the length.

The function to be maximized is $A = lw$.

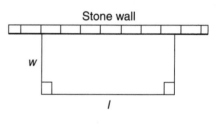

You would prefer a function of a *single* variable. Make the "assumption" that all of the fencing should be used to maximize the area enclosed; then $2w + l = 400$.

$$l = 400 - 2w \Rightarrow A = w(400 - 2w) = 400w - 2w^2, \ 0 < w < 200$$

Always determine the domain. In this problem, you should be clear why the domain is (0, 200): because there are *two* sides of the rectangle with length w and $2w < 400$. (You are limited to 400 ft of fencing.)

Critical numbers: $\dfrac{dA}{dw} = 400 - 4w = 0 \Rightarrow w = 100$ (only critical number)

Max or min? $\dfrac{d^2A}{dw^2} = -4 \Rightarrow \left.\dfrac{d^2A}{dw^2}\right|_{w=100} = -4 < 0$

Therefore, there is a relative maximum at $x = 100$. Since the area function is continuous and it has only one critical number, the relative maximum is the absolute maximum (KEY 28-5A). Hence, the dimensions of the plot with the greatest enclosed area are 100 ft by 200 ft and the largest possible area is 20,000 ft.[2]

• There is an issue you will encounter many times: When not explicitly indicated in the problem, should the boundary points of the domain (in this case, 0 and 200) be included in the domain? Clearly, the area of the rectangle at these endpoints is 0, an absolute minimum. An advantage is that you now have a continuous function on a *closed* interval, and that guarantees the existence of absolute extrema (KEY 28-3). Textbooks have different styles. The bottom line is that your choice does not affect the final solution, so do what you please.

KEY EXAMPLE 29-2 (another rectangle problem)

Rectangle *EASY* with sides parallel to the coordinate axes is inscribed in the region bounded by the graph of $y = -5x^2 + 5$ and the *x*-axis (see the accompanying diagram). Find the coordinates of point *A* in the first quadrant such that the area of *EASY* is maximized.

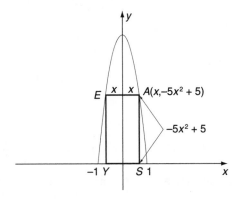

Area function and domain: $A = 2x(-5x^2 + 5) = -10x^3 + 10x$, $0 < x < 1$

Find all critical numbers in the domain: $A'(x) = -30x^2 + 10 = 0 \Rightarrow$

$x = \sqrt{\dfrac{1}{3}}$.

Max or min? $A''(x) = -60x \Rightarrow A''\left(\sqrt{\dfrac{1}{3}}\right) < 0$

Therefore, by the second-derivative test, there is a relative maximum at $x = \sqrt{\dfrac{1}{3}}$. Since the area function is continuous and has only one critical number, the relative maximum is the absolute maximum (KEY 28-5A). Hence, the coordinates of point A that determine the rectangle with the largest area are $\left(\sqrt{\dfrac{1}{3}}, \dfrac{10}{3}\right)$.

KEY EXAMPLE 29-3

Find the maximum volume of a box that can be made by cutting out congruent squares from the corners of an 8-inch by 15-inch rectangular sheet of cardboard and folding up the flaps. (The box is open on top.)

Solution: Draw a diagram, as shown below.

The function: $V = lwh = x(8 - 2x)(15 - 2x) = 4x^3 - 46x^2 + 120x$, $0 < x < 4$

Critical numbers:

$$V'(x) = 12x^2 - 92x + 120 = 4(3x - 5)(x - 6) = 0 \Rightarrow x = \frac{5}{3} \text{ or } x = 6$$

Reject 6 as a critical number since it is not in the domain. The only critical number is $\dfrac{5}{3}$.

$V''(x) = 24x - 92 \Rightarrow V''\left(\dfrac{5}{3}\right) < 0 \Rightarrow$ There is a relative maximum at $x = \dfrac{5}{3}$.

By KEY 28-5, the relative maximum is the absolute maximum. Therefore, the maximum volume is

$$V\left(\frac{5}{3}\right) = \frac{2450}{27} \text{ in.}^3$$

KEY EXAMPLE 29-4 (a practical problem)

A manufacturer wants to make a tuna-fish can with a volume of 7 cubic inches in the shape of a cylinder. To minimize the cost of production, he must minimize the total surface area of the can. What radius and height will minimize the surface area?

Solution: The can is made using three pieces: two circular bases and a rectangular piece that is bent to form the section between the bases. Here are the diagrams:

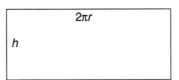

Add the area of the three pieces to get the surface area: $S = 2\pi rh + 2\pi r^2$.

To write S as a function of a single variable, use:

$$V = \pi r^2 h = 7 \Rightarrow h = \frac{7}{\pi r^2}.$$

Therefore, $S = 14r^{-1} + 2\pi r^2$, $r > 0$.

Critical numbers:

$$\frac{dS}{dr} = -14r^{-2} + 4\pi r = 0 \quad \Rightarrow \quad -14 + 4\pi r^3 = 0 \quad \Rightarrow \quad r = \sqrt[3]{\frac{7}{2\pi}}$$

$$\frac{d^2S}{dr^2} = 28r^{-3} + 4\pi = \frac{28}{r^3} + 4\pi \quad \Rightarrow \quad \left.\frac{d^2S}{dr^2}\right|_{r=\sqrt[3]{\frac{7}{2\pi}}} > 0$$

Therefore, at $r = \sqrt[3]{\frac{7}{2\pi}}$ there is a relative minimum. By KEY 28-5B, the relative minimum is the absolute minimum.

$$r = \sqrt[3]{\frac{7}{2\pi}} \quad \Rightarrow \quad h = \frac{7}{\pi r^2} = \frac{7}{\pi \sqrt[3]{\frac{7}{2\pi}}^2} = 2\sqrt[3]{\frac{7}{2\pi}}$$

Hence, the dimensions of a cylindrical can with volume 7 in.^3 and the smallest possible surface area are

$$r = \sqrt[3]{\frac{7}{2\pi}} \approx 2.2 \text{ in.} \quad \text{and} \quad h = 2\sqrt[3]{\frac{7}{2\pi}} \approx 4.4 \text{ in.}$$

NOTE: In this solution, the height is twice the radius. Tuna cans do not have this property. Perhaps product identification and other considerations are more important than minimizing surface area.

Key 30 Motion along a linear path (rectilinear motion)

OVERVIEW *The analysis of the motion of objects is a classic use of calculus. In KEY 20-2, we introduced the relationship between the position and the velocity of a moving object. We devote KEY 30 to revisiting this relationship by posing problems about motion along a line. We also introduce the idea of acceleration.*

KEY REVIEW 30-1

Velocity is the rate of change of **position**. Therefore, the velocity function is the derivative of the position function with respect to time. If the velocity is positive, then the coordinate (position) is increasing and the object is moving in that direction. If the velocity is negative, then the object is moving in the opposite direction. Similarly, **acceleration** is the rate of change of velocity. Therefore, the acceleration function is the derivative of the velocity function (the second derivative of the position function) with respect to time. If the acceleration is positive, then the velocity (while either positive or negative) is increasing. If the acceleration is negative, then the velocity (while either positive or negative) is decreasing.

KEY EXAMPLE 30-2

The distance along a linear path, in light-years, of a flying saucer from some reference point is given by the position function $s(t) = 2t^3 + 10$, $t \geq 0$ (in seconds timed by a stopwatch).

(a) What distance does the saucer travel during the first 3 seconds?
(b) What is its average velocity during the first 3 seconds?
(c) What is its instantaneous velocity after 3 seconds (at $t = 3$)?
(d) What is the average acceleration during the first 3 seconds?
(e) What is its instantaneous acceleration after 3 seconds (at $t = 3$)?

Solutions: Use the given information and the definitions in KEY 30-1; then the position function is $s(t) = 2t^3 + 10$,

the velocity function is $v(t) = \dfrac{ds}{dt} = 6t^2$,

the acceleration function is $a(t) = \dfrac{dv}{dt} = 12t$.

(a) Since the velocity is always positive, the saucer does not change direction. Therefore, the distance it travels during the first 3 sec is the difference of its position at 3 sec and at 0 sec:

$$s(3) - s(0) = 64 - 10 = 54 \text{ light-yr}$$

Note that calculating $s(3)$ gives the saucer's position at $t = 3$, not the distance traveled during the first 3 sec.

(b) Average velocity is the distance traveled divided by the time required to travel that distance:

$$\bar{v}\Big|_{t=0}^{t=3} = \frac{\Delta s}{\Delta t} = \frac{s(3) - s(0)}{3 - 0} = \frac{64 - 10}{3} = 18 \text{ light-yr/sec}$$

(c) Instantaneous velocity is given by the velocity function:

$$v(3) = 54 \text{ light-yr/sec}$$

(d) Average acceleration is the change in velocity divided by the change in time:

$$\bar{a}\Big|_{t=0}^{t=3} = \frac{\Delta v}{\Delta t} = \frac{v(3) - v(0)}{3 - 0} = \frac{54 - 0}{3} = 18 \text{ light-yr/sec/sec}$$

(e) Instantaneous acceleration is given by the acceleration function:

$$a(3) = 36 \text{ light-yr/sec/sec}$$

Key 31 The Mean-Value Theorem

OVERVIEW *The Mean-Value Theorem is one of the centerpieces of elementary differential calculus. Its importance is both practical and theoretical as it is an interesting result that is needed to prove many theorems of calculus. Rolle's Theorem, a special case of the Mean-Value Theorem, is the key to understanding the Mean-Value Theorem.*

KEY THEOREM 31-1 (Rolle's Theorem)

If a function f is continuous on a closed interval $[a, b]$ and differentiable on the open interval (a, b), and if $f(a) = f(b)$, then there exists at least one number c at which $f'(c) = 0$.

- Geometric interpretation: If the graph of a function is continuous (unbroken), differentiable ("smooth"—no cusp points), and $f(a) = f(b)$, then there is at least one point on the graph at which the tangent to the graph is horizontal (slope = 0).
- To get a good feeling for this theorem, you should try to draw a graph that is continuous and smooth so that $f(a) = f(b)$, which does *not* have a horizontal tangent between a and b.

KEY THEOREM 31-2 (Mean-Value Theorem)

If a function f is continuous on a closed interval $[a, b]$ and differentiable on the open interval (a, b), then there exists at least one number c for which $f'(c) = \dfrac{f(b) - f(a)}{b - a}$.

- Geometric interpretation: If the graph of a function is continuous (unbroken) and differentiable ("smooth"—no cusp points), then there is at least one point on the graph at which the tangent to the graph is parallel to the secant line passing through points $(a, f(a))$ and $(b, f(b))$.

KEY DIAGRAM 31-3

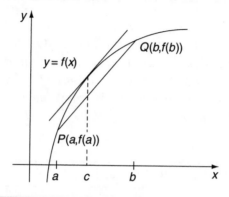

KEY EXAMPLE 31-4

If a car travels 250 miles in 5 hours, then the average speed is 50 miles per hour. But is the car ever traveling *exactly* 50 miles per hour?

Solution: Assume that the car obeys physical laws and that its position function, $s(t)$, is differentiable for $0 < t < 5$ and continuous for $0 \leq t \leq 5$. By the Mean-Value Theorem, there must be an instant, t_1, between 0 hr and 5 hr, when the velocity $v(t_1) = \dfrac{s(5) - s(0)}{5 - 0} = 50$. Therefore, the answer is "yes."

KEY EXAMPLE 31-5

Verify the Mean-Value Theorem for the function $f(x) = x^3$ over the interval $[1, 3]$ by finding the value of c.

Solution:

$$f(x) = x^3 \implies f'(x) = 3x^2$$

$$\frac{f(b) - f(a)}{b - a} = \frac{f(3) - f(1)}{3 - 1} = \frac{27 - 1}{2} = 13$$

$$f'(c) = 3c^2 = 13 \implies c = \pm\sqrt{\frac{13}{3}} \approx \pm 2.082$$

Here, $c \neq -\sqrt{\dfrac{13}{3}}$ because it is not in the interval [1, 3]. Therefore, $c = \sqrt{\dfrac{13}{3}}$.

KEY EXAMPLE 31-6

Does the Mean-Value Theorem "fail" for $f(x) = |x|$ over the interval [−1, 2]? If so, why?

Solution: $\dfrac{f(b) - f(a)}{b - a} = \dfrac{2 - 1}{2 - (-1)} = \dfrac{1}{3}$. But for all x, $f'(x) \neq \dfrac{1}{3}$ (there is no tangent with a slope of $\dfrac{1}{3}$).

The Mean-Value Theorem fails for $f(x) = |x|$ because f is not differentiable at $x = 0$. (See the discussion of KEY 22-1.)

Theme 6 THE INTEGRAL

*I*n Theme 4, we explored the problem of motion. In this theme, we consider a second classic problem: calculating the areas of irregularly shaped regions. Archimedes (287–212 B.C.) found solutions for certain regions using limits. This achievement alone, well before the invention of algebra, qualifies him as one of the world's greatest mathematicians—one who was more than 1500 years ahead of his time! However, without algebra, Archimedes could not find an algorithm for calculating *all* areas. Many centuries passed before Newton and Leibniz found a general solution for all but a "few" areas. Amazingly, their method was related to the problem of motion and used the derivative.

Key 32 Antiderivatives

OVERVIEW *For every process in mathematics there is an **inverse process**, one that undoes the result of the original process. For example, division is the inverse process of multiplication. The inverse of differentiation is **antidifferentiation**. The result of antidifferentiation is an **antiderivative**.*

KEY DEFINITION 32-1

The function g is an **antiderivative** of a function f if and only if $g' = f$.

- The function g is an antiderivative of a function f if and only if f is the derivative of g.
- Note that, while a function cannot have more than one derivative, it can have many antiderivatives. For example, x^3 and $x^3 + 2$ are each antiderivatives of $3x^2$ because $D_x x^3 = 3x^2$ and $D_x(x^3 + 2) = 3x^2$. In fact, for *any* number, C, $x^3 + C$ is an antiderivative of $3x^2$. Therefore, $3x^2$ has an infinite number of antiderivatives.
- In general, if g is an antiderivative of f, then, for every real number C, $g(x) + C$ is an antiderivative of f. In addition, all antiderivatives of f will have the form $g(x) + C$.

KEY DEFINITIONS 32-2

The symbol $\int f(x)\ dx$ represents *all* of the antiderivatives of f, and is called an **indefinite integral**. The process of finding antiderivatives is called **integration** (as well as antidifferentiation).

- When finding derivatives, we merely have to apply several known formulas. However, to find antiderivatives, we need to search for patterns and then apply known formulas.

KEY THEOREM 32-3 (power rule)

$$\int x^n dx = \frac{1}{n+1} x^{n+1} + C, n \neq -1$$

- The case of $n = -1$ is usually discussed in second-semester calculus.

KEY EXAMPLE 32-4

$$\int x^5 dx = \frac{1}{6} x^6 + C$$

KEY THEOREM 32-5

$$\int (f(x) + g(x))\, dx = \int f(x)\, dx + \int g(x)\, dx$$

- "The antiderivative of a sum of functions is the sum of their antiderivatives."

KEY EXAMPLE 32-6

$$\int (x^5 + x^4)\, dx = \int x^5\, dx + \int x^4\, dx = \frac{1}{6} x^6 + \frac{1}{5} x^5 + C$$

KEY THEOREM 32-7

$$\int (f(x) - g(x))\, dx = \int f(x)\, dx - \int g(x)\, dx$$

- "The antiderivative of a difference of functions is the difference of their antiderivatives."

KEY EXAMPLE 32-8

$$\int (x^5 - x^4)\, dx = \int x^5\, dx - \int x^4\, dx = \frac{1}{6}x^6 - \frac{1}{5}x^5 + C$$

KEY THEOREM 32-9

$$\int cf(x)\, dx = c \int f(x)\, dx$$

- "The antiderivative of a constant times a function equals the constant times the antiderivative of the function." (Constant factors can be taken "outside" the integral.)
- It should be clear that, by using the above theorems (KEYS 32-3, 32-5, 32-7, and 32-9), we can integrate any polynomial function.

KEY EXAMPLE 32-10

$$\int 7x^5\, dx = 7 \int x^5\, dx = 7 \cdot \frac{1}{6}x^6 + C = \frac{7}{6}x^6 + C$$

KEY EXAMPLE 32-11

$$\int dx = \int 1\, dx = \int 1x^0\, dx = x + C$$

KEY EXAMPLE 32-12

Find $\int (3x^4 - 8x^3 + 2)\, dx$.

Solution: $\int (3x^4 - 8x^3 + 2)\, dx = \int 3x^4\, dx - \int 8x^3\, dx + \int 2dx$

$$= 3\int x^4\, dx - 8\int x^3\, dx + 2\int 1dx$$

$$= 3\left(\frac{1}{5}x^5\right) - 8\left(\frac{1}{4}x^4\right) + 2(x) + C$$

$$= \frac{3}{5}x^5 - 2x^4 + 2x + C$$

Key 33 Integration by change of variable (substitution)

OVERVIEW *Frequently, we cannot apply simple formulas in order to integrate a function. Sometimes, however, by using a technique called* **change of variable**, *we can simplify the problem so that a formula is applicable. This technique is based on the chain rule in reverse. Changing the variable requires seeing patterns, and that skill is acquired through practice—lots of it!*

KEY EXAMPLE 33-1

Find the derivative of $f(x) = (x^2 + 1)^{10}$.

Solution: By the chain rule,

$$D_x(x^2+1)^{10} = 10(x^2+1)^9(2x) = 20x(x^2+1)^9$$

- A consequence of KEY 33-1 is $\int 20x(x^2+1)^9\,dx = (x^2+1)^{10} + C$. Since we cannot always "see" the antiderivative of a function, we frequently use the *change-of-variable* technique.

KEY EXAMPLE 33-2

Find $\int 20x(x^2 + 1)^9\,dx$.

Solution: Let $u = x^2 + 1$ (the substitution). Then

$$du = 2x\,dx \qquad \text{(differentials, KEY 24-1)}$$

$$\int 20x(x^2+1)^9\,dx = \int 10(x^2+1)^9$$
$$= (2x)\,dx$$
$$= \int 10u^9\,du \qquad \text{(Change the variable.)}$$
$$= u^{10} + C \qquad \text{(power rule)}$$
$$= (x^2+1)^{10} + C \quad \text{(Return to the original variable.)}$$

KEY STRATEGY 33-3

While searching for a substitution that *may* simplify the integration, consider the following possibilities initially:

1. Let *u* equal a function contained in grouping symbols such as parentheses, roots, and fraction lines.
2. Let *u* equal a function whose derivative is a factor of the integrand (the function being integrated).

- Finding a substitution that "will work" requires insight and experience/practice.
- You could expand $(x^2 + 1)^9$ using the binomial expansion theorem and then integrate the resulting polynomial. That method, however, is certainly more tedious and, in addition, is not always applicable (see KEY 33-4).

KEY EXAMPLE 33-4

Find $\int x\sqrt{x^2 + 1}\ dx$.

Solution: Let $u = x^2 + 1 \Rightarrow du = 2x\ dx \Rightarrow x\,dx = \dfrac{1}{2}\ du$. Then

$$\int x\sqrt{x^2+1}\ dx = \frac{1}{2}\int u^{\frac{1}{2}}\ du = \frac{1}{2}\frac{u^{\frac{3}{2}}}{\frac{3}{2}} + C = \frac{1}{2} \cdot \frac{2}{3}u^{\frac{3}{2}} + C$$

$$= \frac{1}{3}(x^2+1)^{\frac{3}{2}} + C = \frac{1}{3}\sqrt{(x^2+1)^3} + C$$

KEY EXAMPLE 33-5

Find $\int \dfrac{7x^2}{(x^3+4)^5}\ dx$.

Solution: Let $u = x^3 + 4 \Rightarrow du = 3x^2\,dx \Rightarrow x^2dx = \dfrac{1}{3}\,du$. Then

$$\int \frac{7x^2}{\left(x^3+4\right)^5}\,dx = 7\int \frac{1}{3}\cdot\frac{1}{u^5}\,du = \frac{7}{3}\int u^{-5}\,du = \frac{7}{3}\cdot\frac{1}{-4}u^{-4} + C$$

$$= -\frac{7}{12}\left(x^3+4\right)^{-4} + C = \frac{-7}{12\left(x^3+4\right)^4} + C$$

- Do not fall into the habit of thinking that a change of variable will always simplify a problem. Look for easier methods first (see KEY 33-6). You must treat each new problem with an open mind. When you are considering a change of variable, make sure that (1) you are able to make the change to the new variable completely and (2) the change simplifies the problem.

KEY EXAMPLE 33-6

Find $\int \dfrac{x+3}{x^5}\,dx$.

Solution: Find this antiderivative *without* changing the variable.

$$\int \frac{x+3}{x^5}\,dx = \int \left(\frac{x}{x^5}+\frac{3}{x^5}\right)dx = \int \left(x^{-4}+3x^{-5}\right)dx$$

$$= -\frac{1}{3}x^{-3} - \frac{3}{4}x^{-4} + C = -\frac{1}{3x^3} - \frac{3}{4x^4} + C$$

KEY EXAMPLE 33-7

Find $\int x\sqrt{2x+1}\,dx$.

Solution: Let $u = 2x+1 \Rightarrow du = 2dx \Rightarrow dx = \dfrac{1}{2}\,du$. Then

$$\int x(2x+1)^{\frac{1}{2}}\,dx = \int \frac{1}{2}(u-1)u^{\frac{1}{2}}\cdot\frac{1}{2}\,du \qquad \left(u = 2x+1 \Rightarrow x = \frac{1}{2}(u-1)\right)$$

$$= \frac{1}{4}\int\left(u^{\frac{3}{2}} - u^{\frac{1}{2}}\right)du \qquad \text{(distributive law)}$$

$$= \frac{1}{4}\left(\frac{2}{5}u^{\frac{5}{2}} - \frac{2}{3}u^{\frac{3}{2}}\right) + C \qquad \text{(power rule)}$$

$$= \frac{1}{10}(2x+1)^{\frac{5}{2}} - \frac{1}{6}(2x+1)^{\frac{3}{2}} + C$$

$$= \frac{1}{10}\sqrt{(2x+1)^5} - \frac{1}{6}\sqrt{(2x+1)^3} + C$$

- When you make a *linear* substitution (e.g., $u = 2x + 1$), it is always possible to completely change to the new variable. Of course, there remains the question, "Does it help?"
- In the preceding example, the linear substitution helped because the expression inside the radical became a monomial $\left(u^{\frac{1}{2}}\right)$ that could be distributed over what became a binomial ($u - 1$), producing only simple powers of the new variable.

KEY EXAMPLE 33-8

Find $\int \sqrt[3]{5x+7}\,dx$.

Solution: Let $u = 5x + 7 \Rightarrow du = 5dx \Rightarrow dx = \frac{1}{5}\,du$. Then

$$\int (5x+7)^{\frac{1}{3}}\,dx = \frac{1}{5}\int u^{\frac{1}{3}}\,du = \frac{1}{5}\cdot\frac{3}{4}u^{\frac{4}{3}} + C$$

$$= \frac{3}{20}(5x+7)^{\frac{4}{3}} + C = \frac{3}{20}\sqrt[3]{(5x+7)^4} + C$$

- Sometimes, the substitution is not at all obvious. As you progress, this will happen with greater frequency. Therefore, it is wise to sharpen your skills now by doing many problems from your textbook.

KEY EXAMPLE 33-9

Find $\int \dfrac{3x-2}{(3x^2-4x)^7}\,dx$.

Solution: Let $u = 3x^2 - 4x \Rightarrow du = (6x-4)dx \Rightarrow (3x-2)dx = \dfrac{1}{2}du$.

[Note the presence of $(3x - 2)\,dx$, which makes the change of variable possible.] Then

$$\int \frac{3x-2}{(3x^2-4x)^7}\,dx = \frac{1}{2}\int \frac{1}{u^7}\,du = \frac{1}{2}\int u^{-7}\,du = \frac{1}{2}\left(-\frac{1}{6}\right)u^{-6} + C$$

$$= -\frac{1}{12}(3x^2-4x)^{-6} + C = \frac{-1}{12(3x^2-4x)^6} + C$$

KEY REMINDERS 33-10

Learn to recognize the patterns that occur frequently. Keep your mind open to new patterns, and try to remember them. A final caution: Be sensitive to possible exceptions to patterns. Dependence on a particular pattern may lead you down a more difficult path. In other words, do not solve problems by rote. THINK! With experience, you will occasionally "see" an antiderivative without changing the variable or using other algebraic procedures.

KEY EXAMPLE 33-11

Find $\int (x-3)^2\,dx$.

Solution: Why change the variable (although it does work)?

$$\int (x-3)^2\,dx = \int (x^2 - 6x + 9)\,dx = \frac{1}{3}x^3 - 3x^2 + 9x + C$$

OR you just may see $\dfrac{1}{3}(x-3)^3 + C$ (easily checked by differentiating using the chain rule).

Key 34 Area and Riemann sums

OVERVIEW *To measure the area of an irregularly shaped region, we use the sum of areas of rectangles as an approximation. Then, reminiscent of instantaneous rate of change, we use limits to get an exact answer. Similar thinking is applicable to measuring other quantities.*

KEY DEFINITIONS 34-1

A **partition** of an interval is a division of the interval into **subintervals**. A **regular partition** is a partition in which all of the subintervals have the same length. A **refinement** of a partition is any new partition that is the result of a division of any subinterval.

- A partition P is frequently defined by its **partition points**: $\{x_0, x_1, x_2, \ldots, x_{n-1}, x_n\}$.

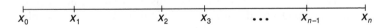

- In a partition, we refer to the first subinterval $[x_0, x_1]$, the second subinterval $[x_1, x_2], \ldots$, the nth subinterval $[x_{n-1}, x_n]$, and, in general, the ith subinterval $[x_{i-1}, x_i]$. The length of the ith subinterval $= \Delta x_i$. The length of the longest subinterval is called the **norm of the partition** and symbolized by $\|P\|$.
- NOTATION: If a function f is continuous on a closed interval, then it is guaranteed to have both maximum and minimum values (KEY 28-3). The maximum value of the function in the ith subinterval is represented by $f_{\max}(x_i)$ and the minimum value by $f_{\min}(x_i)$.

KEY DEFINITIONS 34-2

Given a function $y = f(x)$ that is continuous over some interval I, then the **upper sum** over a given partition is

$$\sum_{i=1}^{n} f_{\max}(x_i)\Delta x_i = f_{\max}(x_1)\Delta x_1 + f_{\max}(x_2)\Delta x_2 + \ldots + f_{\max}(x_n)\Delta x_n$$

and the **lower sum** is

$$\sum_{i=1}^{n} f_{\min}(x_i)\Delta x_i = f_{\min}(x_1)\Delta x_1 + f_{\min}(x_2)\Delta x_2 + \ldots + f_{\min}(x_n)\Delta x_n$$

- If $f(x) \geq 0$, then $f_{\max}(x_i)\,\Delta x_i$ can be thought of as the area of a rectangle with base $= \Delta x_i$ and height $= f_{\max}(x_i)$. The same is true for $f_{\min}(x_i)\,\Delta x_i$.

KEY DIAGRAM 34-3 (an upper sum)

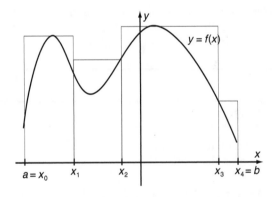

KEY DIAGRAM 34-4 (a lower sum)

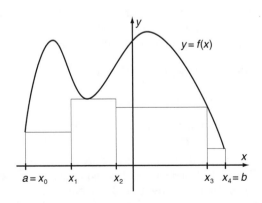

- The upper sum (KEY 34-3) is an approximation (clearly, too large) of the area "under the curve" bounded by $y = f(x)$, $y = 0$ (x-axis), $x = a$, and $x = b$.
- The lower sum (KEY 34-4) is an approximation (too small) of the same region.
- The exact area of the region is between the upper and lower sums.
- As we *refine the partition*, the upper sum "usually" gets smaller and the lower sum "usually" gets larger, with the area always between the two improved approximations (inclusive).

KEY OBSERVATION 34-5

If the function is continuous, then, as the norm of the partition approaches 0 ($\|P\| \to 0$), the upper and lower sums approach the same number, which must be the area of the region.

KEY DEFINITION 34-6

A **Riemann sum** for a function $y = f(x)$ over a partition of some interval $[a, b]$ is *any* sum of the form

$$\sum_{i=1}^{n} f(x_i^*) \Delta x_i = f(x_1^*) \Delta x_1 + f(x_2^*) \Delta x_2 + f(x_3^*) \Delta x_3 + \ldots + f(x_n^*) \Delta x_n$$

where x_i^* is *any* value of x in the ith subinterval.

- For a given partition, the upper and lower sums are the largest and smallest Riemann sums, respectively.
- If the function is continuous, it turns out that

$$\lim_{\|P\| \to 0} (\text{upper sum}) = \lim_{\|P\| \to 0} (\text{lower sum}) = \text{the area}$$

Hence it follows that

$$\lim_{\|P\| \to 0} (\text{any Riemann sum}) = \text{the area}$$

- When writing a Riemann sum, we frequently use regular partitions (in which all subintervals have equal length) and let x_i^* be the left-hand endpoint, the right-hand endpoint, or the midpoint of the subinterval. If the partition is regular, then it should be clear that

$\|P\| \to 0$ is equivalent to $\Delta x \to 0$ and $n \to \infty$ (n = the number of subintervals in the partition).

KEY EXAMPLE 34-7

Evaluate the Riemann sum for the function $f(x) = x^2$ over the interval [0, 2], using the right-hand endpoints of a regular partition with four subintervals.

Solution: The partition points are $\left\{0, \frac{1}{2}, 1, \frac{3}{2}, 2\right\}$, creating four subintervals with equal lengths $\left(\Delta x_i = \frac{1}{2}\right)$. Use right-hand endpoints; then the Riemann sum is

$$\sum_{i=1}^{4} f(x_i^*)\Delta x_i = f\left(\frac{1}{2}\right) \cdot \frac{1}{2} + f(1) \cdot \frac{1}{2} + f\left(\frac{3}{2}\right) \cdot \frac{1}{2} + f(2) \cdot \frac{1}{2}$$

$$= \left(\frac{1}{2}\right)^2 \cdot \frac{1}{2} + \left(\frac{2}{2}\right)^2 \cdot \frac{1}{2} + \left(\frac{3}{2}\right)^2 \cdot \frac{1}{2} + \left(\frac{4}{2}\right)^2 \cdot \frac{1}{2}$$

$$= \frac{1}{8} + \frac{4}{8} + \frac{9}{8} + \frac{16}{8} = \frac{30}{8} = 3.75$$

- The Riemann sum is an approximation of the area under the graph of $f(x) = x^2$ from $x = 0$ to $x = 2$. In KEY 34-8, you will find that the exact area of the same region is $\frac{8}{3} \approx 2.667$.
- Draw the graph of $f(x) = x^2$ from $x = 0$ to $x = 2$, and include the rectangles used in this Riemann sum. Could you have predicted that the approximation would be larger than the actual area?

KEY EXAMPLE 34-8

Evaluate the Riemann sum for the function $f(x) = x^2$ over the interval [0, 2], using the right-hand endpoints of a regular partition with eight subintervals. (This is the same region as in KEY 34-7, but with a refinement of the partition.)

Solution: The partition points are $\left\{0, \frac{1}{4}, \frac{1}{2}, \frac{3}{4}, 1, \frac{5}{4}, \frac{3}{2}, \frac{7}{4}, 2\right\}$, creating eight subintervals with equal lengths $\left(\Delta x_i = \dfrac{1}{4}\right)$. Use right-hand endpoints; then the Riemann sum is

$$\sum_{i=1}^{8} f(x_i^*)\Delta x_i = \left(\frac{1}{4}\right)^2 \cdot \frac{1}{4} + \left(\frac{2}{4}\right)^2 \cdot \frac{1}{4} + \left(\frac{3}{4}\right)^2 \cdot \frac{1}{4} + \left(\frac{4}{4}\right)^2 \cdot \frac{1}{4} + \left(\frac{5}{4}\right)^2 \cdot \frac{1}{4}$$

$$+ \left(\frac{6}{4}\right)^2 \cdot \frac{1}{4} + \left(\frac{7}{4}\right)^2 \cdot \frac{1}{4} + \left(\frac{8}{4}\right)^2 \cdot \frac{1}{4}$$

$$= \frac{1}{64} + \frac{4}{64} + \frac{9}{64} + \frac{16}{64} + \frac{25}{64} + \frac{36}{64} + \frac{49}{64} + \frac{64}{64}$$

$$= \frac{204}{64} = 3.1875$$

- Compare the results in KEYS 34-7 and 34-8 with the actual area. Note how a refinement of a partition resulted in a better approximation.
- In KEY 34-7, $\Delta x = 0.5$; and in KEY 34-8, $\Delta x = 0.25$. Generally, the closer Δx is to 0, the better the approximation. If you evaluate $\lim\limits_{\Delta x \to 0}$ (Riemann sum), letting the norm of the partition approach 0, you get the exact area!

Key 35   The definite integral and notation

OVERVIEW *The number that all Riemann sums approach as the norm of the partition approaches 0 is so important that it is given a name: the **definite integral**.*

KEY NOTATION 35-1

Given a function f and an interval $[a, b]$, if there is a number that all Riemann sums approach as $\|P\| \to 0$, it is represented by $\int_a^b f(x)\, dx$, which is read as "the definite integral from a to b of the function f."

- If f is a positive-valued function, then $\int_a^b f(x)\, dx =$ "the area under the graph from a to b."
- The definite integral has many parts, all of which remind us of the limit of Riemann sums.
    1. The numbers a and b are the endpoints of the interval and are called the **limits of integration**.
    2. The integral sign, $\int$, resembles an elongated S, which reminds us that we are taking the limit of a *S*um.
    3. The expressions dx and $f(x)$ remind us, respectively, of the base (Δx_i) and the height $(f(x_i^*))$ of rectangles; $f(x)$ is called the **integrand**.
- Although $\int_a^b f(x)\, dx$ and $\int f(x)\, dx$ are similar, they represent entirely different things.
    1. $\int f(x)\, dx$ represents all of the antiderivatives of f, a family of *functions*, and is called an **indefinite integral**.
    2. $\int_a^b f(x)\, dx$ represents a *number* and is called a **definite integral**.
    3. The notations are similar because it turns out that they are related. In KEY 35-2, we shall learn how to evaluate a definite integral using antidifferentiation instead of a tedious evaluation of the *limit* of Riemann sums.

KEY DEFINITION 35-2

If the definite integral $\int_a^b f(x)\, dx$ (a limit) exists, then the function is said to be **integrable** on the interval $[a, b]$.

- Since the limit may not exist, there is the complex problem of determining which functions are integrable. That problem is beyond the scope of this book, but KEY 35-3 helps.

KEY THEOREM 35-3

IF (a) f is continuous on $[a, b]$ **OR** (b) f is bounded ($-B \leq f(x) \leq B$ for some $B > 0$) and has a finite number of discontinuities on $[a, b]$, **THEN** f is integrable on $[a, b]$.

KEY OBSERVATION 35-4

Consider a function f whose graph is shown below.

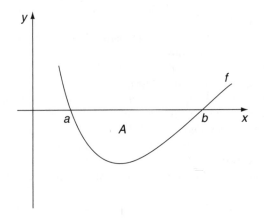

Since $f(x) < 0$ on $[a, b]$, all Riemann sums will be negative. Therefore, $A = \int_a^b f(x)\, dx < 0$. Since area cannot be negative, the area between the graph and the x-axis from a to b is $|A| = -\int_a^b f(x)\, dx$.

KEY EXAMPLE 35-5

What is the total area between the graph of h and the x-axis shown below?

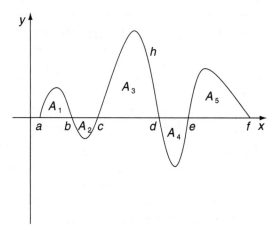

Solution: The total area between the graph of h and the x-axis is

$$|A_1| + |A_2| + |A_3| + |A_4| + |A_5| = \int_a^f |h(x)| dx$$

See KEYS 37-5, 40-5, 40-6, 40-7, 40-8, and 46-2 for related discussions.

- **WARNING:** Frequently, $|\int_a^f h(x)\ dx| \neq \int_a^f |h(x)|\ dx$. To understand why, consider the following:

$$|(1) + (-2) + (3) + (-4) + (5)| \neq |1| + |-2| + |3| + |-4| + |5|$$

Key 36   Two fundamental theorems

OVERVIEW *The evaluation of a definite integral is very important. However, its evaluation by taking the limit of Riemann sums as the norm of the partition approaches 0 is laborious. This procedure is the generalized solution for calculating areas that Archimedes, lacking algebra, could not find. It took the brilliant insights of Newton and Leibniz to see the connection between the limit of Riemann sums and differentiation (or, more accurately, antidifferentiation). This connection greatly simplifies the evaluation of a definite integral.*

KEY THEOREM 36-1

$$\int_a^a f(x)\,dx = 0$$

- Intuitively, this should make good sense: the area under the curve is 0 if the length of the interval is 0.

KEY THEOREM 36-2

$$\int_b^a f(x)\,dx = -\int_a^b f(x)\,dx$$

- "If we integrate from right to left, then we get the additive inverse of the result obtained by integrating from left to right."

KEY THEOREM 36-3 (the additive property of integrals)

$$\int_a^b f(x)\,dx + \int_b^c f(x)\,dx = \int_a^c f(x)\,dx$$

- Intuitively, the area under the curve from a to b plus the area under the curve from b to c equals the area under the curve from a to c.

KEY THEOREM 36-4 (the First Fundamental Theorem of calculus)

If a function is continuous on an interval, then it has antiderivatives on the interval.

- IMPORTANT NOTE: The existence of antiderivatives is proved by showing that the function F, defined by $F(x) = \int_a^x f(t)\, dt$, is an antiderivative of f (t is a dummy variable: KEY 22-5). Therefore, $D_x(\int_a^x f(t)\, dt) = f(x)$. Recalling that a is a constant, we see that $\int_a^x f(t)\, dt$ is a function of x. When we evaluate the definite integral using the Second Fundamental Theorem (KEY 36-5), we substitute a and x into an antiderivative, getting an algebraic expression in terms of x and constants.
- The powerful First Fundamental Theorem asserts that continuity guarantees the existence of antiderivatives. However, it does not tell us *how* to find them. That task may even be impossible.

KEY THEOREM 36-5 (the Second Fundamental Theorem of calculus)

$\int_a^b f(x)\, dx = g(b) - g(a)$, where g is any antiderivative of f.

- **To evaluate a definite integral:** Find *any* antiderivative of the function, evaluate it at the endpoints of the interval, and subtract the two results in the correct order. (WOW!)

KEY EXAMPLE 36-6

Evaluate: $\int_0^3 4x\, dx$.

Solution: First, find an antiderivative of $f(x) = 4x$. By KEYS 32-3 and 32-9, one antiderivative is $g(x) = 2x^2$. Then, by the Second Fundamental Theorem,

$$\int_0^3 4x\, dx = g(3) - g(0) = 2(3)^2 - 2(0)^2 = 18$$

- Look at the accompanying graph of $f(x) = 4x$.

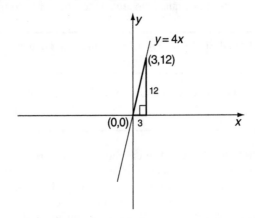

- Note that the area "under the graph" (but above the x-axis) between $x = 0$ and $x = 3$ is a right triangle whose area is 18.

KEY NOTE 36-7

If a function f is continuous and nonnegative, then, for $a < b$,

$$\int_a^b f(x)\,dx = \text{the area under the curve}$$

- KEY 37 has more examples.

Key 37   Evaluating definite integrals

OVERVIEW *In this key, we apply the two fundamental theorems to evaluate definite integrals.*

KEY NOTATION 37-1

$g(x)\vert_a^b$ means $g(b) - g(a)$.

KEY EXAMPLE 37-2

Evaluate: $\int_0^2 x^2\, dx$.

Solution:

$$\int_0^2 x^2\, dx = \frac{1}{3}x^3\Big|_0^2 \qquad \text{(Use KEY 32-3 to find an antiderivative.)}$$

$$= \frac{1}{3}(2^3 - 0^3) = \frac{8}{3} \quad \text{(See the discussion of KEY 34-7.)}$$

- This method is much simpler than evaluating the limits of Riemann sums.
- When finding the antiderivative requires a change of variable, you have a choice of two methods:
    1. Find an antiderivative, return to the original variable, and use the original limits of integration.
    2. Change the limits of integration as you change the variable.

KEY EXAMPLE 37-3

Find the area under the graph of $f(x) = x\sqrt{x^2+1}$ from $x = \sqrt{3}$ to $x = \sqrt{15}$.

Solution: Evaluate $\displaystyle\int_{\sqrt{3}}^{\sqrt{15}} x\sqrt{x^2+1}\, dx$.

Let $u = x^2 + 1 \Rightarrow du = 2x\, dx \Rightarrow x\, dx = \dfrac{1}{2}\, du$. Then

$$\int_{\sqrt{3}}^{\sqrt{15}} x\sqrt{x^2+1}\, dx = \frac{1}{2}\int_4^{16} u^{\frac{1}{2}}\, du \quad \begin{pmatrix} x = \sqrt{3} \;\Rightarrow u = 4 \\ x = \sqrt{15} \Rightarrow u = 16 \end{pmatrix}$$

$$= \frac{1}{2}\cdot\frac{2}{3} u^{\frac{3}{2}}\Big|_4^{16} \quad \text{(KEY 32-3, the power rule)}$$

$$= \frac{1}{3}\left(16^{\frac{3}{2}} - 4^{\frac{3}{2}}\right) = \frac{1}{3}(64 - 8) = \frac{56}{3}$$

KEY NOTE 37-4

In general, if the substitution is $u = g(x)$,

$$\int_a^b f(x)\, dx = \int_{g(a)}^{g(b)} h(u)\, du$$

- **WARNING:** When you change the variable, be sure that the substitution is continuous over the interval of integration (it usually is).
- In KEY 37-3, since $u = x^2 + 1$ is continuous (it is polynomial) over $[\sqrt{3}, \sqrt{15}]$, the substitution is permissible.

KEY EXAMPLE 37-5

Evaluate: $\int_{-2}^{6} |x|\, dx$.

Solution: The problem is that no formulas are available for integrating the absolute value function. You cannot "remove" the absolute value, however, unless you know whether x is positive or negative. The additive property of integrals (KEY 36-3) to the rescue!

$$\int_{-2}^{6} |x|\, dx = \int_{-2}^{0} |x|\, dx + \int_{0}^{6} |x|\, dx \qquad \text{(KEY 36-3)}$$

$$= \int_{-2}^{0} (-x)\, dx + \int_{0}^{6} x\, dx \qquad \text{(KEY 2-8)}$$

$$= -\frac{1}{2} x^2 \Big|_{-2}^{0} + \frac{1}{2} x^2 \Big|_{0}^{6} \qquad \text{(KEYS 32-3 and 36-5)}$$

$$= (0 - (-2)) + (18 - 0)$$

$$= 20$$

- The choice of intervals $[-2, 0]$ and $[0, 6]$ is based on where x is positive and where x is negative.
- Draw the graph of $y = |x|$. Notice that you could have calculated the area under the curve from $x = -2$ to $x = 6$ by adding the areas of two right triangles. Needless to say, this option is not usually available.

Key 38   The average value of a
function

OVERVIEW *The method of calculating the average of a finite number of data is well known. A natural question arises: Is it possible to find the average of an infinite number of data? In this key, we define how to do this in an intuitively appealing way. This definition demonstrates the relevance of calculus to statistics.*

KEY OBSERVATION 38-1

Consider a function, $y = f(x)$, that is continuous on a closed interval $[a, b]$. In KEY 35-1, we summarized a method for calculating the area under the curve. Surely, there must be some rectangle, whose base is the interval, that has the exact same area. What would be the height of that rectangle?

KEY DIAGRAM 38-2

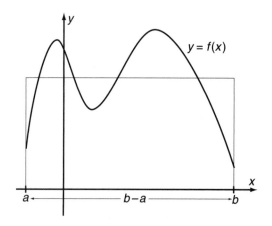

- Here is the answer to the question posed in KEY 38-1: In order to have the same area, the height of the rectangle must be the *average* height (average value) of the function. Note in the diagram that the length of the base is $b - a$, so the area under the curve is equal to both $(b - a)f_{avg}$ and $\int_a^b f(x)\,dx$.

KEY DEFINITION 38-3

If a function is continuous on the interval $[a, b]$, then the average value of the function over that interval is given by

$$f_{avg} = \frac{1}{b-a} \int_a^b f(x)\,dx$$

- Continuity and the Intermediate-Value Theorem guarantee that the function achieves its average value at least once in the interval.
- Although our explanation has been limited to positive-valued functions because the situation is easy to visualize, the definition should also make sense for functions that have negative values.

KEY EXAMPLE 38-4

Find (a) the average value of the function $y = x^2$ over the interval $[1, 4]$, and (b) all values of x in the interval at which the function reaches its average value.

Solutions: (a) $f_{avg} = \frac{1}{4-1} \int_1^4 x^2\,dx = \frac{1}{3} \cdot \frac{1}{3} x^3 \Big|_1^4 = \frac{1}{9}(64 - 1) = 7$

(b) $x^2 = 7 \Rightarrow x = \sqrt{7}$ or $x = -\sqrt{7}$. Reject $-\sqrt{7}$ because it is not in the interval. Therefore, the function reaches its average value only at $x = \sqrt{7}$.

KEY EXAMPLE 38-5

Consider the function

$$f(x) = \begin{cases} 1 & \text{if } 1 \le x < 3 \\ 2 & \text{if } 3 \le x \le 5 \end{cases}$$

The average value of f over the interval $[1, 5]$ is 1.5. Clearly, the function never reaches its average value. How can that happen?

Solution: There is a point of discontinuity at $x = 3$. Since the function is not continuous on the interval, the Intermediate-Value Theorem does not apply.

Key 39 Trapezoidal rule;
Simpson's rule

OVERVIEW *Sometimes, it is extremely difficult, even impossible, to evaluate a definite integral. In these circumstances, we must settle for approximations. Riemann sums use the areas of rectangles with horizontal tops. If we use tops that are chords, then the approximating figures will be trapezoids. The chords more closely conform to the shape of the graph of the function and, therefore, yield a better approximation for the same partition. Perhaps, if we use parabolic segments to approximate the graph, the results will be even more accurate. There are formulas for calculating both types of approximation.*

KEY DIAGRAM 39-1

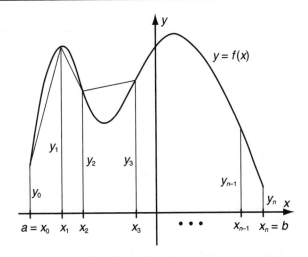

KEY FORMULA 39-2

For a given function f and a *regular* partition of the interval $[a, b]$, the trapezoidal approximation to the area under the curve is given by

$$T = \Delta x \left(\frac{1}{2} y_0 + y_1 + y_2 + \ldots + y_{n-1} + \frac{1}{2} y_n \right)$$

where $y_i = f(x_i)$ for $i = 0, 1, 2, \ldots, n$.

KEY EXAMPLE 39-3

Evaluate the trapezoidal approximation for the function $f(x) = x^2$ over the interval $[0, 2]$ using a regular partition with four subintervals.

Solution: The partition points are $\left\{ 0, \frac{1}{2}, 1, \frac{3}{2}, 2 \right\}$, creating four subintervals with equal lengths $\left(\Delta x = \frac{1}{2} \right)$.

$$T = \frac{1}{2} \left(\frac{1}{2} \cdot 0^2 + \left(\frac{1}{2} \right)^2 + 1^2 + \left(\frac{3}{2} \right)^2 + \frac{1}{2} \cdot 2^2 \right) = \frac{11}{4} = 2.75$$

KEY FORMULA 39-4

For a given function f and a *regular* partition with an *even* number of subintervals of the interval $[a, b]$, Simpson's approximation to the area under the curve is given by

$$S = \frac{\Delta x}{3} (y_0 + 4y_1 + 2y_2 + 4y_3 + 2y_4 + \ldots + 2y_{n-2} + 4y_{n-1} + y_n)$$

KEY EXAMPLE 39-5

Evaluate Simpson's approximation for the function $f(x) = x^2$ over the interval $[0, 2]$ using a regular partition with four subintervals.

Solution: The partition points are $\left\{ 0, \frac{1}{2}, 1, \frac{3}{2}, 2 \right\}$, creating four subintervals with equal lengths $\left(\Delta x = \frac{1}{2} \right)$.

$$S = \frac{1}{3} \cdot \frac{1}{2} \left(0^2 + 4\left(\frac{1}{2}\right)^2 + 2(1)^2 + 4\left(\frac{3}{2}\right)^2 + 2^2 \right) = \frac{8}{3} \approx 2.667$$

- We have approximated the area under $y = x^2$ from $x = 0$ and $x = 2$ using four different methods: upper sum, lower sum, trapezoids, and Simpson's rule. We now compare their accuracy.

Upper sum = $\frac{45}{12}$, lower sum = $\frac{21}{12}$, trapezoids = $\frac{33}{12}$, Simpson's rule = $\frac{32}{12}$. The exact area = $\frac{32}{12}$. Notice how much more accurate the trapezoids were as compared to upper and lower sums. Of course, there exists a Riemann sum exactly equal to the area, but finding it is "impossible." The reason that Simpson's rule was perfectly accurate in this particular example is that this method uses parabolic segments to approximate the graph and $y = x^2$ is a parabola, so the fit was perfect.

Theme 7 APPLICATIONS OF
INTEGRATION

We have seen that the definite integral, as the limit of approximations, can be used to evaluate areas of irregularly shaped regions. We now explore how to approximate other types of geometric measures and use the definite integral to obtain exact answers.

Key 40 The area between two curves

OVERVIEW *In theme 6, we restricted our discussion to areas of regions bounded by the graph of one positive-valued function and the **x**-axis between two values of **x** (two vertical lines: **x** = **a** and **x** = **b**). We now consider areas bounded by two curves and two lines parallel to one of the coordinate axes.*

KEY DIAGRAM 40-1

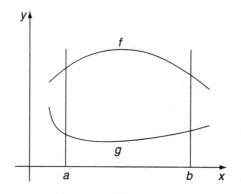

KEY FORMULA 40-2

The area bounded by f and g between $x = a$ and $x = b$ as in KEY 40-1 is

$$A = \int_a^b [f(x) - g(x)]\, dx$$

- Since subtraction is not commutative, it is important to determine which function is larger.
- If the limits of integration are not supplied, then they must be implicit in the problem (e.g., points of intersection).
- This formula applies regardless of the positions of f and g relative to the x-axis (above and/or below).

KEY EXAMPLE 40-3

Find the area bounded by $y = x^2 + x$ and $y = x^2 + 1$ from $x = 2$ to $x = 4$.

Solution: Note that $x^2 + x \geq x^2 + 1$ for all $x \in [2, 4]$.

$$A = \int_2^4 [(x^2 + x) - (x^2 + 1)]\, dx = \int_2^4 (x - 1)\, dx$$

$$= \left(\frac{1}{2} x^2 - x \right)\Big|_2^4 = 4 - 0 = 4$$

KEY EXAMPLE 40-4

Find the area enclosed by $y = x^2 - 3$ and $y = 5 - x^2$.

Solution: No limits of integration are given, so find the points of intersection:

$$x^2 - 3 = 5 - x^2 \Rightarrow 2x^2 = 8 \Rightarrow x = \pm 2$$

Now determine which function is larger by trying *one* value of x in the interval of integration. For $x = 0$, $5 - x^2 \geq x^2 - 3$. Therefore

$$A = \int_{-2}^2 [(5 - x^2) - (x^2 - 3)]\, dx = \int_{-2}^2 (8 - 2x^2)\, dx$$

$$= \left(8x - \frac{2}{3} x^3 \right)\Big|_{-2}^2 = \left(16 - \frac{16}{3} \right) - \left(-16 + \frac{16}{3} \right) = \frac{64}{3}$$

- Because the boundaries, $y = x^2 - 3$ and $y = 5 - x^2$, are symmetric to the y-axis, so is the region. You could have found the area from $x = 0$ to $x = 2$ and doubled the result $\left(\text{i.e., } A = 2\int_0^2 [(5 - x^2) - (x^2 - 3)]\, dx \right)$. This method would slightly reduce the number of required calculations.
- If there are *more than two points of intersection*, then, between each pair of *consecutive* points of intersection, determine which function is larger. Then partition the interval of integration at the points of intersection and use the additive property of integrals (KEY 36-3). The variation in KEY 40-5 of the formula in KEY 40-2 includes all of these observations.

KEY FORMULA 40-5

$$A = \int_a^b |f(x) - g(x)|\, dx$$

- Before integrating, you need to remove the absolute-value signs. This can be done only if you know in which intervals $f(x) \geq g(x)$ and in which intervals $g(x) \geq f(x)$. You can review a previous example in KEY 37-5.

KEY DIAGRAM 40-6

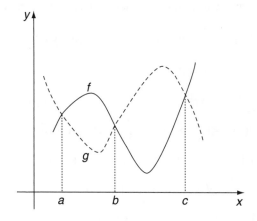

- Note that $f(x) > g(x)$ in (a, b) and $g(x) > f(x)$ in (b, c). The area enclosed by the graphs of f and g is

$$A = \int_a^c |f(x) - g(x)|\, dx = \int_a^b |f(x) - g(x)|\, dx + \int_b^c |f(x) - g(x)|\, dx$$
$$= \int_a^b [f(x) - g(x)]\, dx + \int_b^c [g(x) - f(x)]\, dx$$

KEY EXAMPLE 40-7

Find the area of the region enclosed by $y = 2x^3 - 3x^2 + 8x + 1$ and $y = x^3 + 4x^2 - 6x + 9$.

Solution: First find the points of intersection:

$$2x^3 - 3x^2 + 8x + 1 = x^3 + 4x^2 - 6x + 9$$
$$x^3 - 7x^2 + 14x - 8 = 0$$

Use the rational-root, remainder, and factor theorems to find that the points of intersection are at $x = 1$, $x = 2$, and $x = 4$. Therefore

$$(2x^3 - 3x^2 + 8x + 1) - (x^3 + 4x^2 - 6x + 9) = (x - 1)(x - 2)(x - 4)$$

Now do a number-line analysis.

After determining which function is larger in each subinterval, calculate:

$$A = \int_1^4 |(2x^3 - 3x^2 + 8x + 1) - (x^3 + 4x^2 - 6x + 9)| \, dx$$

$$= \int_1^2 |(2x^3 - 3x^2 + 8x + 1) - (x^3 + 4x^2 - 6x + 9)| \, dx$$

$$+ \int_2^4 |(2x^3 - 3x^2 + 8x + 1) - (x^3 + 4x^2 - 6x + 9)| \, dx$$

$$= \int_1^2 [(2x^3 - 3x^2 + 8x + 1) - (x^3 + 4x^2 - 6x + 9)] \, dx$$

$$+ \int_2^4 [(x^3 + 4x^2 - 6x + 9) - (2x^3 - 3x^2 + 8x + 1)] \, dx$$

$$= \int_1^2 (x^3 - 7x^2 + 14x - 8) \, dx + \int_2^4 (-x^3 + 7x^2 - 14x + 8) \, dx$$

$$= \left(\frac{1}{4}x^4 - \frac{7}{3}x^3 + 7x^2 - 8x \right) \Big|_1^2 + \left(-\frac{1}{4}x^4 + \frac{7}{3}x^3 - 7x^2 + 8x \right) \Big|_2^4$$

$$= \left[\left(4 - \frac{56}{3} + 28 - 16 \right) - \left(\frac{1}{4} - \frac{7}{3} + 7 - 8 \right) \right]$$

$$+ \left[\left(-64 + \frac{448}{3} - 112 + 32 \right) - \left(-4 + \frac{56}{3} - 28 + 16 \right) \right]$$

$$= \frac{5}{12} + \frac{8}{3} = \frac{37}{12}$$

KEY EXAMPLE 40-8

Find the area between $y = x^3$ and the x-axis from $x = -2$ to $x = 2$.

Solution: $A = \int_{-2}^{2} x^3 \, dx = \left(\frac{1}{4} x^4 \right)\Big|_{-2}^{2} = \frac{1}{4}(16 - 16) = 0$

But this cannot be correct! What went wrong? In the interval of integration, the function $y = x^3$ is both above and below the x-axis. The area below the x-axis is counted as negative in the integral. To calculate the total area, you must consider a function with regions above the x-axis but congruent to the original regions. The function $y = |x^3|$ has that property, so you need to evaluate

$$A = \int_{-2}^{2} |x^3| \, dx = \int_{-2}^{0} |x^3| \, dx + \int_{0}^{2} |x^3| \, dx$$

$$= \int_{-2}^{0} (-x^3) \, dx + \int_{0}^{2} (x^3) \, dx$$

$$= \left(-\frac{1}{4} x^4 \right)\Big|_{-2}^{0} + \left(\frac{1}{4} x^4 \right)\Big|_{0}^{2}$$

$$= [0 - (-4)] + [4 - 0] = 8$$

KEY EXAMPLE 40-9

Find the area bounded by $y = x^2$, $y = 4$, and $x = 0$.

Solution 1:

$A = \int_{0}^{2} (4 - x^2) \, dx$

$= \left(4x - \frac{1}{3} x^3 \right)\Big|_{0}^{2}$

$= \left(8 - \frac{8}{3} \right) - 0 = \frac{16}{3}$

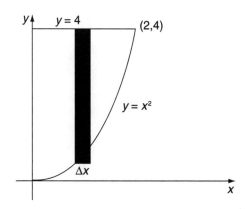

Solution 2:

$$A = \int_0^4 \left(\sqrt{y} - 0\right) dy = \int_0^4 y^{\frac{1}{2}} \, dy$$

$$= \left(\frac{2}{3} x^{\frac{3}{2}}\right)\Bigg|_0^4 = \frac{2}{3}(8-0) = \frac{16}{3}$$

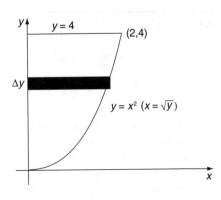

- The graphs of $y = x^2$ and $y = 4$ intersect at $(2,4)$. It should be clear why the limits of integration are different in the two solutions.

KEY OBSERVATION 40-10

It is very important to appreciate the possibility of two approaches. In Solution 1, the "thickness" of the approximating rectangle is Δx, and we integrated with respect to x. In Solution 2, the thickness of the rectangle is Δy, and we integrated with respect to y. In this example, both methods were equally easy. In others, one approach may be significantly easier than the other. These choices will recur when we calculate volumes, arc lengths, and surface areas.

Key 41 Volumes in general by slicing

OVERVIEW *We now consider a general method for calculating the volume of a solid. The formula is derived by slicing the solid (like a loaf of bread) and then adding the volumes of the slices. The sum of these volumes is a Riemann sum, and its limit is a definite integral.*

KEY FORMULA 41-1

Suppose that a solid is positioned relative to a v number line (usually the x-axis or y-axis) and its extremities are at $v = a$ and $v = b$. If, when "slicing" the solid with a plane perpendicular to the number line, it is possible to express the area of the *cross section* as a continuous function $A(v)$ of the variable v, then the volume of the solid is

$$V = \int_a^b A(v)\, dv$$

KEY EXAMPLE 41-2

The base of a certain solid is the circle $x^2 + y^2 = 25$. If the solid is sliced with a plane perpendicular to the **x-axis**, the cross section is an isosceles right triangle with one of its legs being a chord of the circle. What is the volume of the solid?

Solution:

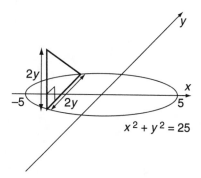

The area of an isosceles right triangle is one-half the product of the lengths of the (congruent) legs. The length of each leg is $2y$, where y is the second coordinate of the point on the circular base for the value of x at which the slicing takes place. Then, by KEY 41-1,

$$V = \int_{-5}^{5} \frac{1}{2}(2y)^2 \, dx = 2\int_{-5}^{5} y^2 \, dx = 2\int_{-5}^{5} (25 - x^2) \, dx$$

$$= 2\left(25x - \frac{1}{3}x^3\right)\Big|_{-5}^{5} = 2\left[\left(125 - \frac{125}{3}\right) - \left(-125 + \frac{125}{3}\right)\right]$$

$$= 2\left[250 - \frac{250}{3}\right] = \frac{1000}{3}$$

Key 42   Volumes of revolution by disks and washers

OVERVIEW *If a region in the **xy**-plane is revolved around a line (the "line of revolution") parallel to one of the axes, the solid generated is called a **volume of revolution**. Using KEY 41-1, we can derive formulas for calculating the volumes of these special solids.*

KEY DIAGRAM 42-1

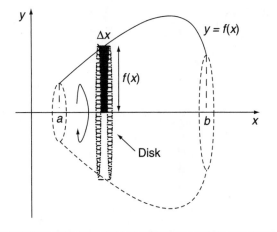

KEY FORMULA 42-2

A region bounded by a function, $y = f(x)$, the x-axis, and two lines perpendicular to the x-axis ($x = a$ and $x = b$) is revolved about the x-axis, producing a solid of revolution. If the solid is sliced with a plane perpendicular to the x-axis, the cross section is a circle. By KEY 41-1, the volume of the solid is

$$V = \pi \int_a^b r^2 \, dx = \pi \int_a^b f^2(x) \, dx$$

- To visualize the *circular* cross-section, it may help to imagine a partition of the interval on the x-axis and a "typical" approximating rectangle (like that used for a Riemann sum approximation to the area of a region) revolving about the x-axis, creating a "thin," circular cylinder called a **disk**.
- Similar reasoning applies if the region bounded by $x = g(y)$, the y-axis, $y = c$, and $y = d$ is revolved about the y-axis. Then the volume is

$$V = \pi \int_c^d r^2 \, dy = \pi \int_c^d g^2(y) \, dy$$

KEY EXAMPLE 42-3

The region bounded by $y = x^2 + 1$, $y = 0$, $x = 1$, and $x = 2$ is revolved about the x-axis. Find the volume of the solid formed.

Solution: By disks

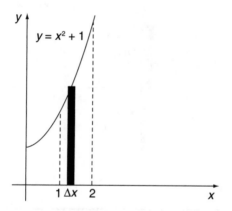

The thickness of the rectangle is Δx, so integrate with respect to x (KEY 40-10). As the rectangle revolves about the x-axis, it creates a *disk*. Use KEY 42-2; then

$$V = \pi \int_1^2 (x^2 + 1)^2 \, dx = \pi \int_1^2 (x^4 + 2x^2 + 1) \, dx$$

$$= \pi \left(\frac{1}{5} x^5 + \frac{2}{3} x^3 + x \right) \Big|_1^2$$

$$= \pi \left[\left(\frac{32}{5} + \frac{16}{3} + 2 \right) - \left(\frac{1}{5} + \frac{2}{3} + 1 \right) \right] = \frac{178\pi}{15}$$

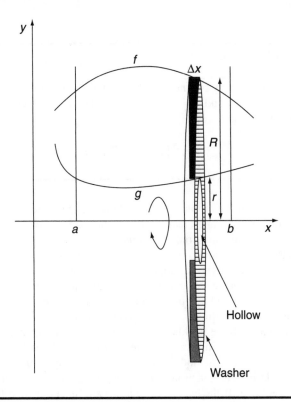

- If the region shown in KEY 42-4 is revolved about the x-axis, imagine the typical rectangle revolving about the x-axis as well. The rectangle creates a hollow disk, called a **washer**. The cross section, shown below, is the region between two concentric circles.

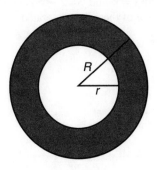

The area between the two concentric circles is $A = \pi R^2 - \pi r^2$, where R is the measure of the "outer radius" and r is the measure of the "inner radius."

KEY FORMULA 42-5

If a region bounded by two nonnegative functions, $y = f(x)$ and $y = g(x)$, and the lines $x = a$ and $x = b$, with $f(x) \geq g(x)$, is revolved about the x-axis, then the volume is

$$V = \pi \int_a^b \left[f^2(x) - g^2(x) \right] dx$$

- No less important than the formula is the ability to visualize the revolving rectangle creating the washer and to express the outer radius and the inner radius as functions of x.
- If the cross sections are perpendicular to the y-axis, then all of the preceding observations apply to the variable y.
- When solving a volume problem, draw a reasonably accurate sketch. (A graphing calculator will help, if one is available.) Recall that the thickness of the rectangle dictates the variable of integration. If the thickness is Δx, integrate with respect to x. If the thickness is Δy, integrate with respect to y (KEY 40-10).
- Although this method is not always so simple, if the *curved* boundaries are functions of x, try setting up an integral with respect to x. If the *curved* boundaries are functions of y, try integrating with respect to y.

KEY EXAMPLE 42-6

The region enclosed by $y = x^2$ and $y = 2 - x^2$ is revolved about the x-axis ($y = 0$). Find the volume of the solid generated.

Solution: By washers

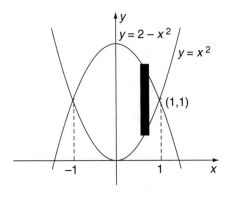

$$V = \pi \int_{-1}^{1} \left[(2 - x^2)^2 - (x^2)^2 \right] dx$$

$$= \pi \int_{-1}^{1} (4 - 4x^2) \, dx$$

$$= \pi \left(4x - \frac{4}{3} x^3 \right) \Big|_{-1}^{1}$$

$$= \pi \left[\left(4 - \frac{4}{3} \right) - \left(-4 + \frac{4}{3} \right) \right] = \frac{16\pi}{3}$$

KEY EXAMPLE 42-7

The region enclosed by $y = x^2$ and $y = 2 - x^2$ is revolved about $y = -1$. Find the volume of the solid generated.

Solution: By washers

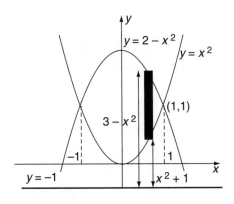

$$V = \pi \int_{-1}^{1} \left[\left(2 - x^2 + 1\right)^2 - \left(x^2 + 1\right)^2 \right] dx$$

$$= \pi \int_{-1}^{1} \left[\left(9 - 6x^2 + x^4\right) - \left(x^4 + 2x^2 + 1\right) \right] dx$$

$$= \pi \int_{-1}^{1} \left(8 - 8x^2\right) dx = 8\pi \int_{-1}^{1} \left(1 - x^2\right) dx$$

$$= 8\pi \left(x - \frac{1}{3} x^3 \right) \Big|_{-1}^{1} = 8\pi \left[\left(1 - \frac{1}{3}\right) - \left(-1 + \frac{1}{3}\right) \right] = \frac{32\pi}{3}$$

• Compare KEYS 42-6 and 42-7. Note that a change in the line of revolution produces different outer and inner radii.

KEY EXAMPLE 42-8

The region in the first quadrant bounded by $y = x^2$, $y = 2 - x^2$, and $y = 0$ is revolved about the x-axis. Find the volume of the solid of revolution.

Solution: In this problem, the lengths of the radii of the *disks* are determined by different functions of x in different parts of the domain. Therefore, add the volumes of the solids generated by the regions to the left and right of $x = 1$.

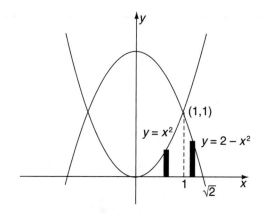

$$V = \pi \int_0^1 (x^2)^2 \, dx + \pi \int_1^{\sqrt{2}} (2 - x^2)^2 \, dx = \pi \int_0^1 x^4 dx + \pi \int_1^{\sqrt{2}} (4 - 2x^2 + x^4) \, dx$$

$$= \pi \left(\frac{1}{5} x^5 \right) \Big|_0^1 + \pi \left(4x - \frac{2}{3} x^3 + \frac{1}{5} x^5 \right) \Big|_1^{\sqrt{2}}$$

$$= \pi \left(\frac{1}{5} - 0 \right) + \pi \left[\left(4\sqrt{2} - \frac{4}{3} \sqrt{2} + \frac{4}{5} \sqrt{2} \right) - \left(4 - \frac{2}{3} + \frac{1}{5} \right) \right]$$

$$= \frac{52\sqrt{2} - 50}{15} \pi$$

KEY EXAMPLE 42-9

The region bounded by $x = 3 - 2y - y^2$ and $x = 0$ is revolved about $x = 0$. Find the volume of the solid of revolution.

Solution: Since the boundaries are functions of y, prepare to integrate with respect to y. First, find the points of intersection:

$$3 - 2y - y^2 = 0 \quad \Rightarrow \quad y = -3 \text{ or } y = 1$$

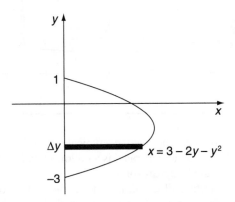

As the rectangle (with thickness Δy) is revolved about the y-axis, it generates a *disk*. Then

$$V = \pi \int_{-3}^1 (3 - 2y - y^2)^2 \, dy = \pi \int_{-3}^1 (9 - 12y - 2y^2 + 4y^3 + y^4) \, dy$$

$$= \pi \left(9y - 6y^2 - \frac{2}{3} y^3 + y^4 + \frac{1}{5} y^5 \right) \Big|_{-3}^1 = \frac{26\pi}{15}$$

Key 43 Volumes of revolution by cylindrical shells

OVERVIEW *In KEY 42, we considered regions bounded by functions of* **x** *(functions of* **y***) that were revolved about lines parallel to the* **x***-axis (***y***-axis). How does the situation change if the region is bounded by functions of one variable and is revolved about a line parallel to the "opposite" axis?*

KEY DIAGRAM 43-1

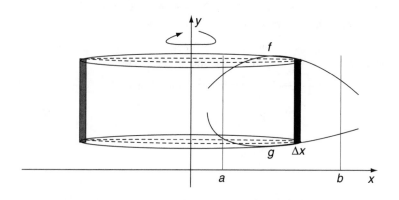

- In the above diagram, imagine a typical rectangle revolving about the *y*-axis. The solid generated is a hollow *cylinder* with thickness Δx, that is, a **cylindrical shell**.
- While the surface of a disk is a circle, the surface of a shell is a cylinder. The lateral surface area of a cylinder is $S = 2\pi r h$.

KEY FORMULA 43-2

When calculating volumes using cylindrical shells,

$$V = 2\pi \int_a^b r h \, dv$$

where dv = the thickness of the rectangle. Whether it is Δx or Δy will determine the variable of integration. Also, h = the height of the rectangle expressed as a function of the variable of integration, and r = the radius of revolution (the distance of the rectangle from the line of revolution) expressed as a function of the variable of integration.

KEY EXAMPLE 43-3

The region bounded by $y = x^2$ and $y = 4$ is revolved about the y-axis. Find the volume of the solid generated.

Solution 1: By shells

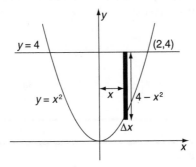

Since the thickness of the rectangle is Δx, integrate with respect to x. Then

$$V = 2\pi \int_0^2 x \overbrace{(4 - x^2)}^{h} \, dx = 2\pi \int_0^2 (4x - x^3) \, dx$$

$$= 2\pi \left(2x^2 - \frac{1}{4}x^4 \right) \Big|_0^2$$

$$= 2\pi(4 - 0) = 8\pi$$

- The limits of integration are *not* −2 and 2 because the "half-area" to the left of the y-axis generates the *exact same solid* as the half-area to the right, and you do not want to count the volume of the solid twice!
- *If* you revolve the region about the x-axis, then the two half-areas generate *congruent but different* solids and you *will* integrate from $x = -2$ to $x = 2$.
- In this problem, you could have easily integrated with respect to y.

Solution 2: By disks

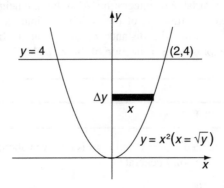

Now the thickness is Δy, so integrate with respect to y. As the rectangle revolves around the y-axis, it generates *disks*. Therefore

$$V = \pi \int_0^4 x^2 \, dy = \pi \int_0^4 y \, dy$$

$$= \pi \left(\frac{1}{2} y^2 \right) \Big|_0^4 = \pi(8 - 0) = 8\pi$$

KEY EXAMPLE 43-4

The region in the first quadrant bounded by $y = x^2$, $y = 4$, and $x = 0$ is revolved about $x = -1$. Find the volume of the solid generated.

Solution 1: Using rectangles with thickness Δx

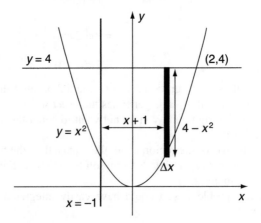

Revolving about $x = -1$, the typical rectangle generates *cylindrical shells*. Therefore

$$V = 2\pi \int_0^2 (x+1)(4-x^2)\,dx = 2\pi \int_0^2 (-x^3 - x^2 + 4x + 4)\,dx$$

$$= 2\pi \left(-\frac{1}{4}x^4 - \frac{1}{3}x^3 + 2x^2 + 4x \right)\Bigg|_0^2$$

$$= 2\pi \left[\left(-4 - \frac{8}{3} + 8 + 8 \right) - 0 \right] = \frac{56\pi}{3}$$

- Note that the radius of revolution is $x + 1$, the distance of the typical rectangle from $x = -1$, the line of revolution.

Solution 2: Using rectangles with thickness Δy

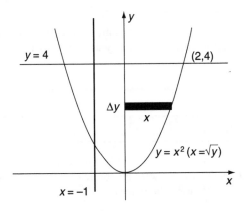

Revolving about $x = -1$, the rectangle generates *washers*. The outer radius is $x + 1$, and the inner radius is 1. Then

$$V = \pi \int_0^4 \left[(x+1)^2 - 1^2 \right] dy = \pi \int_0^4 \left[(\sqrt{y}+1)^2 - 1^2 \right] dy = \pi \int_0^4 \left(y + y^{\frac{1}{2}} \right) dy$$

$$= \pi \left(\frac{1}{2}y^2 + \frac{4}{3}y^{\frac{3}{2}} \right)\Bigg|_0^4 = \pi \left[\left(8 + \frac{32}{3} \right) - 0 \right] = \frac{56\pi}{3}$$

- In solution 2, V was first expressed as a definite integral in the form $\int_a^b g(x)\,dy$. Such "mixing" of the variables in the integral during intermediate steps sometimes helps to clarify the solution process. *However, be sure to express the integral in terms of one variable before evaluating it.*

Key 44 Parametric equations; the
length of an arc

OVERVIEW *Consider the graph of a differentiable function (continuity is not enough). Using a definite integral, we can find the length of the graph (called the **arc length**) between any two points on the graph. First, however, we review **parametric equations**.*

KEY REVIEW 44-1

In the context of this discussion, **parametric equations** are two equations that define the variables x and y as functions of a third variable (usually t).

KEY EXAMPLE 44-2

Let $x = t + 1$ and $y = 4 - t^2$. These are two parametric equations defining x and y as functions of t. A useful algebraic exercise is to express y as a function of x ("eliminating the parameter").

$$x = t + 1 \implies t = x - 1$$
$$y = 4 - t^2 \implies y = 4 - (x-1)^2 = -x^2 + 2x + 3$$

- Every function $y = f(x)$ can be defined using parameters, and parametric equations are very useful in science. For example, suppose that these equations define the parabolic path, $y = -x^2 + 2x + 3$, of a rocket that is to land on the Moon. It is not enough to know that the paths of the rocket and the Moon intersect; they must reach the point of intersection at the same time. The parametric equations not only define the path of the rocket but also include its location on the path at any instant (as a function of time).
- When eliminating the parameter (as with any algebraic procedure), it is imperative to maintain the original restrictions on the variable. For example, if $x = \cos^2 t$ and $y = \sin^2 t$, then $x \geq 0$ and

$y \geq 0$. After we eliminate the parameter t, we have $x + y = 1$, for which x or y could be negative. We should write $x + y = 1$, $x \geq 0$, and $y \geq 0$.

Now we consider the measurement of **arc length**.

KEY DIAGRAM 44-3

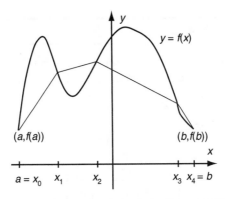

- By using chords in each subinterval of a partition as approximations of different parts of the arc length, we have a Riemann sum. If we let the norm of the partition approach 0 (KEY 34-1), the sum approaches a definite integral (KEY 35-1). Depending on certain algebraic steps in the derivation (which may be found in most calculus textbooks), we can derive three different formulas for arc length.

KEY FORMULAS 44-4

The length, s, of a graph between two points is

$$s = \begin{cases} \int_a^b \sqrt{1 + \left(\dfrac{dy}{dx}\right)^2}\, dx & (1) \\[3mm] \int_c^d \sqrt{\left(\dfrac{dx}{dy}\right)^2 + 1}\, dy & (2) \\[3mm] \int_e^f \sqrt{\left(\dfrac{dx}{dt}\right)^2 + \left(\dfrac{dy}{dt}\right)^2}\, dt & (3) \end{cases}$$

- Use equation (1) if y is a function of x; use equation (2) if x is a function of y; use equation (3) if x and y are functions of a parameter (t).
- There will be times when more than one of the above conditions is true. In that case, you will have a choice of formula. The decision may be critical: because of the presence of the square root, one formula may be significantly more difficult to use than another.
- Of course, the limits of integration are for the variable of integration.

KEY NOTATION 44-5

Many books write the arc-length formula as $s = \int ds$, where ds represents an increment of arc defined by

$$ds = \begin{cases} \sqrt{1 + \left(\dfrac{dy}{dx}\right)^2} \; dx \\[2ex] \sqrt{\left(\dfrac{dx}{dy}\right)^2 + 1} \; dy \\[2ex] \sqrt{\left(\dfrac{dx}{dt}\right)^2 + \left(\dfrac{dy}{dt}\right)^2} \; dt \end{cases}$$

These are equivalent formulas.

KEY EXAMPLE 44-6

Find the length of the graph of $y = \dfrac{2}{3} x^{\frac{3}{2}}$ between points (0,0) and (9,18).

Solution: Since y is a function of x, use KEY 44-4(1). Then

$$y = \frac{2}{3} x^{\frac{3}{2}} \quad \Rightarrow \quad \frac{dy}{dx} = x^{\frac{1}{2}} \quad \Rightarrow \quad \left(\frac{dy}{dx}\right)^2 = x$$

$$s = \int_a^b \sqrt{1 + \left(\frac{dy}{dx}\right)^2} \; dx = \int_0^9 \sqrt{1+x} \; dx = \int_0^9 (1+x)^{\frac{1}{2}} \; dx$$

$$(u = 1+x \quad \Rightarrow \quad du = dx)$$

$$= \int_1^{10} u^{\frac{1}{2}} du = \left(\frac{2}{3} u^{\frac{3}{2}}\right)\Bigg|_1^{10} = \frac{2}{3}\left(10^{\frac{3}{2}} - 1^{\frac{3}{2}}\right) = \frac{2}{3}\left(10\sqrt{10} - 1\right)$$

KEY EXAMPLE 44-7

Find the arc length of $x = \dfrac{y^3}{6} + \dfrac{1}{2y}$ from $y = 1$ to $y = 2$.

Solution: Since x is a function of y, use KEY 44-4(2). Then

$$x = \frac{1}{6}y^3 + \frac{1}{2}y^{-1} \Rightarrow \frac{dx}{dy} = \frac{1}{2}y^2 - \frac{1}{2}y^{-2} \Rightarrow \left(\frac{dx}{dy}\right)^2 = \frac{1}{4}y^4 - \frac{1}{2} + \frac{1}{4}y^{-4}$$

$$s = \int_a^b \sqrt{\left(\frac{dx}{dy}\right)^2 + 1}\, dy = \int_1^2 \sqrt{\frac{1}{4}y^4 + \frac{1}{2} + \frac{1}{4}y^{-4}}\, dy$$

$$= \int_1^2 \sqrt{\left(\frac{1}{2}y^2 - \frac{1}{2}y^{-2}\right)^2}\, dy = \int_1^2 \left|\frac{1}{2}y^2 - \frac{1}{2}y^{-2}\right|\, dy \quad \text{(KEY 2-9C)}$$

$$= \int_1^2 \left(\frac{1}{2}y^2 - \frac{1}{2}y^{-2}\right) dy \qquad\qquad \left(\frac{1}{2}y^2 - \frac{1}{2}y^{-2} > 0\right)$$

$$= \left(\frac{1}{6}y^3 - \frac{1}{2}y^{-1}\right)\Bigg|_1^2 = \left(\frac{4}{3} - \frac{1}{4}\right) - \left(\frac{1}{6} - \frac{1}{2}\right) = \frac{17}{12}$$

KEY EXAMPLE 44-8

Find the arc length of the graph defined by the parametric equations $x = 4t - 2$ and $y = 3t + 1$ for $0 \leq t \leq 2$.

Solution 1: $s = \int_a^b \sqrt{\left(\dfrac{dx}{dt}\right)^2 + \left(\dfrac{dy}{dt}\right)^2}\, dt = \int_0^2 \sqrt{4^2 - 3^2}\, dt = \int_0^2 5\, dt$

$$= (5t)\big|_0^2 = 10 - 0 = 10$$

Solution 2: Eliminate the parameter, then

$$t = \frac{1}{4}(x + 2) \quad \Rightarrow \quad y = \frac{3}{4}x + \frac{5}{2}$$

This is the equation of a straight line, so you can simply use the distance formula to calculate the distance between $(-2, 1)$ and $(6, 7)$. (The coordinates of the points are found by using the parametric equations with $t = 0$ and $t = 2$.)

$$s = \sqrt{(x_2 - x_1)^2 + (y_2 - y_1)^2} = \sqrt{(-2 - 6)^2 + (1 - 7)^2} = \sqrt{(-8)^2 + (-6)^2} = 10$$

Key 45 The area of a surface of revolution

OVERVIEW *If an arc of a graph is revolved about a line, it creates a* **surface of revolution**. *The definite integral is used to calculate the area of this surface. The line of revolution will usually be parallel to one of the axes.*

KEY DIAGRAM 45-1

Suppose an arc is revolved about the *y*-axis, for example. Again using chords to approximate the graph, imagine the arc revolving about the *y*-axis.

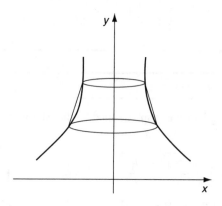

The chord, as it revolves about the line, creates a surface called a frustum of a cone. The surface-area formula is derived from this approximation.

KEY FORMULA 45-2

If an arc is revolved about a line, then the surface area is

$$S = 2\pi \int_a^b r \, ds$$

where r = the radius of revolution (expressed as a function of the variable of integration) and ds = one of the three alternative forms described in KEY 44-5.

If the line of revolution is	then the radius of revolution is
the y-axis	$r = x$
the x-axis	$r = y$
$x = -2$	$r = x + 2$
$y = -2$	$r = y + 2$
$x = 3$	$r = x - 3$
etc.	

KEY EXAMPLE 45-3

The curve $y = \sqrt[3]{x}$ from $(1,1)$ to $(8,2)$ is revolved about the y-axis. Find the area of the surface generated.

Solution 1: Integrating with respect to y

$$y = x^{\frac{1}{3}} \Rightarrow x = y^3 \Rightarrow \frac{dx}{dy} = 3y^2 \Rightarrow \left(\frac{dx}{dy}\right)^2 = 9y^4$$

$$S = 2\pi \int_a^b r\, ds = 2\pi \int_1^2 x\sqrt{\left(\frac{dx}{dy}\right)^2 + 1}\, dy = 2\pi \int_1^2 y^3 \sqrt{9y^4 + 1}\, dy$$

$$\left(u = 9y^4 + 1 \Rightarrow du = 36y^3\, dy \Rightarrow \frac{1}{36} du = y^3\, dy\right)$$

$$= \frac{\pi}{18} \int_{10}^{145} u^{\frac{1}{2}} du = \frac{\pi}{18} \left(\frac{2}{3} u^{\frac{3}{2}}\right)\Big|_{10}^{145} = \frac{\pi}{27}\left(145\sqrt{145} - 10\sqrt{10}\right)$$

Solution 2: Integrating with respect to x

$$y = x^{\frac{1}{3}} \Rightarrow \frac{dy}{dx} = \frac{1}{3} x^{-\frac{2}{3}} \Rightarrow \left(\frac{dy}{dx}\right)^2 = \frac{1}{9} x^{-\frac{4}{3}}$$

$$S = 2\pi \int_a^b r\, ds = 2\pi \int_1^8 x\sqrt{1 + \left(\frac{dy}{dx}\right)^2}\, dx$$

$$= 2\pi \int_1^8 x\sqrt{1 + \frac{1}{9} x^{-\frac{4}{3}}}\, dx = 2\pi \int_1^8 x\sqrt{\frac{1}{9} x^{-\frac{4}{3}}\left(9x^{\frac{4}{3}} + 1\right)}\, dx$$

$$= 2\pi \int_1^8 x \left(\frac{1}{3}x^{-\frac{2}{3}}\right)\sqrt{9x^{\frac{4}{3}}+1}\ dx = \frac{2\pi}{3}\int_1^8 x^{\frac{1}{3}}\left(9x^{\frac{4}{3}}+1\right)^{\frac{1}{2}} dx$$

$$\left(u = 9x^{\frac{4}{3}}+1 \Rightarrow du = 12x^{\frac{1}{3}}\ dx \Rightarrow \frac{1}{12}du = x^{\frac{1}{3}}\ dx\right)$$

$$= \frac{\pi}{18}\int_{10}^{145} u^{\frac{1}{2}}du = \frac{\pi}{18}\left(\frac{2}{3}u^{\frac{3}{2}}\right)\Bigg|_{10}^{145} = \frac{\pi}{27}\left(145\sqrt{145}-10\sqrt{10}\right)$$

- In this problem, there is a choice. Since y is given as a function of x, the first instinct is to integrate with respect to x (and that is frequently wise). However, in this case, integrating with respect to y is a bit easier. Your ability to *anticipate* where each choice leads will help you to make the "correct" choice initially.

KEY EXAMPLE 45-4

The arc of $y = \sqrt{a^2 - x^2}$, $a > 0$, $-a \le x \le a$, is revolved about the x-axis. Find the surface area generated.

Solution:

$$y = (a^2 - x^2)^{\frac{1}{2}} \Rightarrow \frac{dy}{dx} = \frac{1}{2}(a^2-x^2)^{-\frac{1}{2}}(-2x) \Rightarrow \left(\frac{dy}{dx}\right)^2 = \frac{x^2}{a^2-x^2}$$

$$S = 2\pi\int_a^b r\,ds = 2\pi\int_{-a}^a y\sqrt{1+\frac{x^2}{a^2-x^2}}\ dx = 2\pi\int_{-a}^a \sqrt{a^2-x^2}\sqrt{1+\frac{x^2}{a^2-x^2}}\ dx$$

$$= 2\pi\int_{-a}^a \sqrt{a^2-x^2+x^2}\,dx = 2\pi\int_{-a}^a \sqrt{a^2}\,dx = 2\pi a\int_{-a}^a dx$$

$$= 2\pi a(x)\big|_{-a}^a = 2\pi a[a-(-a)] = 4\pi a^2$$

- Here, $y = \sqrt{a^2 - x^2}$ is the upper semicircle of $x^2 + y^2 = a^2$, the circle with center at the origin and radius $= a$. When this is revolved about the x-axis, it generates a sphere. The result obtained above verifies that the surface area of a sphere with radius $= a$ is $S = 4\pi a^2$.

KEY EXAMPLE 45-5

The arc of $x = \dfrac{y^4}{8} + \dfrac{1}{4y^2}$ from $y = 1$ to $y = 2$ is revolved about the line $y = -1$. Find the area of the surface generated.

Solution: The algebra in this problem is very similar to that in KEY 44-7.

$$x = \frac{1}{8}y^4 + \frac{1}{4}y^{-2} \Rightarrow \frac{dx}{dy} = \frac{1}{2}y^3 - \frac{1}{2}y^{-3} \Rightarrow \left(\frac{dx}{dy}\right)^2 = \frac{1}{4}y^6 - \frac{1}{2} + \frac{1}{4}y^{-6}$$

$$S = 2\pi \int_a^b r\, ds = 2\pi \int_1^2 (y+1) \sqrt{\left(\frac{dx}{dy}\right)^2 + 1}\; dy$$

$$= 2\pi \int_1^2 (y+1) \sqrt{\frac{1}{4}y^6 + \frac{1}{2} + \frac{1}{4}y^{-6}}\; dy = 2\pi \int_1^2 (y+1) \sqrt{\left(\frac{1}{2}y^3 + \frac{1}{2}y^{-3}\right)^2}\; dy$$

$$= 2\pi \int_2^1 (y+1) \left| \frac{1}{2}y^3 + \frac{1}{2}y^{-3} \right| dy$$

$$= 2\pi \int_1^2 (y+1) \left(\frac{1}{2}y^3 + \frac{1}{2}y^{-3}\right) dy \quad \text{NOTE: } \frac{1}{2}y^3 + \frac{1}{2}y^{-3} > 0 \text{ for } 1 \le y \le 2.$$

$$= \pi \int_1^2 (y+1)(y^3 + y^{-3})\, dy = \pi \int_1^2 (y^4 + y^3 + y^{-2} + y^{-3})\, dy$$

$$= \pi \left(\frac{1}{5}y^5 + \frac{1}{4}y^4 - y^{-1} - \frac{1}{2}y^{-2} \right) \Bigg|_1^2$$

$$= \pi \left[\left(\frac{32}{5} + 4 - \frac{1}{2} - \frac{1}{2} \right) - \left(\frac{1}{5} + \frac{1}{4} - 1 - \frac{1}{2} \right) \right] = \frac{209\pi}{20}$$

Key 46 Motion along a path (revisited)

OVERVIEW *In KEY 30 we found that the derivative of the position function with respect to time is the velocity function and the derivative of the velocity function with respect to time is the acceleration function. Now we reason backward: using integration, we can find the position function from the velocity function, and the velocity function from the acceleration function.*

KEY FORMULAS 46-1

For an object traveling along a path:

A. Its position function $s(t) = \int v(t)\, dt$

B. Its velocity function $v(t) = \int a(t)\, dt$

KEY FORMULA 46-2

The *total distance* traveled between $t = t_1$ and $t = t_2$ is

$$d = \int_{t_1}^{t_2} |v(t)|\, dt$$

- It is important to understand the reason for the absolute value in the formula. For example, suppose the object is moving vertically. Then $v(t) > 0$ means that the object is rising (the distance from the ground is increasing) and $v(t) < 0$ means that the object is falling (the distance from the ground is decreasing). If the object is both rising and falling during the time interval $[t_1, t_2]$ and finishes where it started, then $\int_{t_1}^{t_2} v(t)\, dt = 0$. This result is caused by an accumulation of "positive" and "negative" distances. Nevertheless, the object may have traveled a considerable distance in moving up and down. The absolute value guarantees that all motion accumulates positive distance. (Analogous situations can be found in KEYS 37-5, 40-5, and 40-6.)

- Many problems deal with objects moving near Earth's surface. If gravity is the only force acting on the object, then the acceleration function is $a(t) = -32$ feet per second per second (ft/sec/sec) or $a(t) = -9.8$ meters per second per second (m/s/s), depending on the units of measurement.

KEY EXAMPLE 46-3

An arrow is shot upward from a cliff 980 meters above the ground with a velocity of 49 meters per second.

(a) What is the maximum height attained by the arrow?
(b) With what velocity does the arrow pass the archer on the way down?
(c) From the time it is shot, how long does the arrow take to hit the ground at the bottom of the cliff?
(d) What is the speed of the arrow when it hits the ground?

Solutions: From the given information, you can find the velocity and position functions.

$a(t) = -9.8$	(constant negative acceleration due to gravity)
$v(t) = -9.8t + C_1$	(Velocity is *an* antiderivative of acceleration.)
$v(0) = 49$	(called an *initial* or *boundary condition*)

$$v(0) = 49 \quad \Rightarrow \quad C_1 = 49 \Rightarrow \quad \boxed{v(t) = -9.8t + 49}$$

$s(t) = -4.9t^2 + 49t + C_2$ (Position is *an* antiderivative of velocity.)
$s(0) = 980$ (another initial condition)

$$s(0) = 980 \quad \Rightarrow \quad C_2 = 980 \quad \Rightarrow \quad \boxed{s(t) = -4.9t^2 + 49t + 980}$$

(a) Since $a(t) = s''(t) < 0$ for all t, any critical number for $s(t)$ must be a maximum (KEY 27-6). Also, at the maximum height, the velocity is 0 (the arrow is neither rising nor falling), so

$$v(t) = -9.8t + 49 = 0 \quad \Rightarrow \quad t = 5$$

To find the maximum height, evaluate $s(5) = 1102.5$. The maximum height attained by the arrow is 1102.5 m.

(b) To find the velocity of the arrow as it passes the archer, you need to know *when* the arrow passes the archer. That occurs when $s(t) = 980$.

$$-4.9t^2 + 49t + 980 = 980$$
$$-4.9t^2 + 49t = 0 \quad \Rightarrow \quad t(-4.9t + 49) = 0 \quad \Rightarrow \quad t = 0 \text{ or } t = 10$$

Reject $t = 0$ because that is the time when the archer shot the arrow. To find the velocity at $t = 10$, evaluate $v(10) = -49$. The velocity of the arrow as it passes the archer is -49 m/s/s. The velocity is negative because the arrow is falling.

(c) The arrow hits the ground when $s(t) = 0$.

$$s(t) = -4.9t^2 + 49t + 980 = 0$$
$$t^2 - 10t - 200 = 0 \quad \text{(Divide both sides of the equation by } -4.9.)$$
$$(t + 10)(t - 20) = 0 \Rightarrow t = -10 \text{ or } t = 20$$

Reject $t = -10$; hence the arrow takes 20 s to hit the ground.

(d) To find the speed of the arrow when it hits the ground, first evaluate $v(20) = -147$. The speed of the arrow is 147 m/s when it hits the ground.

- The term *speed* means "how fast regardless of direction" and is always nonnegative. Technically, velocity is a vector and speed is a scalar: speed $= |\text{velocity}|$.

KEY EXAMPLE 46-4

A particle moves along a number line with a velocity given by $v(t) = t^2 - 6t + 8$, from $t = 0$ to $t = 5$. Find (a) the total distance traveled and (b) the net displacement. Position is measured in millimeters, and time in seconds.

Solutions:

(a) Total distance $= \int_a^b |v(t)|\, dt = \int_0^5 |t^2 - 6t + 8|\, dt$. To remove the absolute value (so you can integrate), use a number line to analyze when $t^2 - 6t + 8$ is positive and when negative.

194

$$d = \int_0^5 \left| t^2 - 6t + 8 \right| dt$$

$$= \int_0^2 \left| t^2 - 6t + 8 \right| dt + \int_2^4 \left| t^2 - 6t + 8 \right| dt + \int_4^5 \left| t^2 - 6t + 8 \right| dt$$

$$= \int_0^2 \left(t^2 - 6t + 8 \right) dt + \int_2^4 \left(-t^2 + 6t - 8 \right) dt + \int_4^5 \left(t^2 - 6t + 8 \right) dt$$

$$= \left(\frac{1}{3} t^3 - 3t^2 + 8t \right) \Big|_0^2 + \left(-\frac{1}{3} t^3 + 3t^2 - 8t \right) \Big|_2^4 + \left(\frac{1}{3} t^3 - 3t^2 + 8t \right) \Big|_4^5$$

$$= \frac{20}{3} + \frac{4}{3} + \frac{4}{3} = \frac{28}{3}$$

The total distance traveled by the particle in the given time interval is $\frac{28}{3}$ mm.

(b) Net displacement $= \int_a^b v(t)\, dt = \int_0^5 \left(t^2 - 6t + 8 \right) dt$

$$= \left(\frac{1}{3} t^3 - 3t^2 + 8t \right) \Big|_0^5 = \frac{20}{3} - 0 = \frac{20}{3}$$

During the given time interval, the particle moved 20/3 mm to the right.

• From part (a), you could calculate that the particle moved a total of 8 mm to the right and $\frac{4}{3}$ mm to the left.

Key 47   A physical application: work

OVERVIEW *In physics,* **work** *is defined as force times distance, where the force exerted on an object in the direction of motion is constant. What happens if the force is variable? Integration to the rescue!*

KEY FORMULA 47-1

If an object moves a distance d due to a constant force F in the direction of motion, then the work done by the force is

$$W = Fd$$

- If the force is measured in pounds and the distance in feet, then the unit of work is the foot-pound (ft-lb). In the metric system, common units of work are the newton-meter (also called a joule) and the dyne-centimeter (erg).
- The force required to lift an object is its weight.

KEY EXAMPLE 47-2

If an object moves 15 feet under a *constant* force of 10 pounds, the work exerted is 150 foot-pounds.

KEY FORMULA 47-3

An object moves along a number line from $x = a$ to $x = b$ under a variable force that is a continuous function, $F(x)$, of its position, x. By partitioning the interval $[a, b]$ and adding the work done in each subinterval, we have a Riemann sum. Taking the limit as the norm of the partition approaches 0, we can derive that the work exerted by the force is

$$W = \int_a^b F(x)dx$$

- A perfect example of this situation is Hooke's Law.

KEY FORMULA 47-4 (Hooke's Law)

When a spring is stretched (or compressed) x units from its *normal length*, the force it exerts is proportional to x. The force function is $F(x) = kx$, where k is the spring constant, which varies from spring to spring.

KEY EXAMPLE 47-5

A spring has a normal length of 30 inches. A force of 6 pounds is required to stretch the spring 12 inches beyond its normal length. How much work would be required to stretch the spring to a length of 50 inches?

Solution: First, find the spring constant. By Hooke's Law, $F(x) = kx$. When $x = 12$, $F(x) = 6$. Hence

$$6 = 12k \Rightarrow k = \frac{1}{2} \Rightarrow F(x) = \frac{1}{2}x$$

Then

$$W = \int_0^{20} \frac{1}{2}x \, dx = \left(\frac{1}{4}x^2 \right)\Bigg|_0^{20} = 100 \text{ in.-lb}$$

- NOTE: The limits of integration are 0 and 20 because x represents the amount of stretching beyond the normal length of the spring: from 30 inches (0 inches beyond normal length) to 50 inches (20 inches beyond normal length).

KEY EXAMPLE 47-6

Thirty tons of fuel are loaded onto a rocket that weighs 4 tons. As the rocket climbs, it burns fuel at the constant rate of 2 tons for each 1000 feet of ascent. How much work is required to lift the rocket to an altitude of 4000 feet? (Assume that the force of gravity is the same constant at different altitudes.)

Solution: The force required to lift the rocket is its weight (in tons), which varies because, as fuel burns off, the rocket loses weight. Let $x =$ the altitude of the rocket. Then the force function is

$$F(x) = 34 - 2\left(\frac{x}{1000}\right)$$

Note that the weight is 34 tons when $x = 0$ (the rocket is on the ground). As x increases, $\frac{x}{1000}$ represents the part of 1000 ft to which the rocket has been raised, and $2\left(\frac{x}{1000}\right)$ is the weight of the burned fuel.

$$F(x) = 34 - 2\left(\frac{x}{1000}\right) = 34 - \frac{1}{500}x$$

$$W = \int_0^{4000} \left(34 - \frac{1}{500}x\right) dx = \left(34x - \frac{1}{1000}x^2\right)\Big|_0^{4000}$$

$$= (136{,}000 - 16{,}000) - (0) = 120{,}000 \text{ ft-tons} = 240{,}000{,}000 \text{ ft-lb}$$

KEY EXAMPLE 47-7

A cylindrical tank with radius 10 feet and height 40 feet is half-filled with water that weighs 62.4 pounds per cubic foot. How much work is required to pump all the water over the upper lip of the tank?

Solution: First divide (partition) the water into n thin slabs, each slab being a short cylinder, as shown in the accompanying diagram.

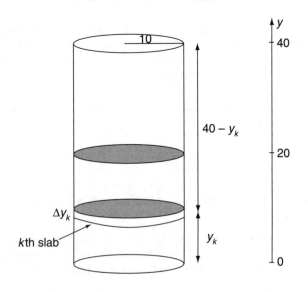

The force required to "lift" the kth slab is its weight. Recall that the volume of a cylinder is $V = \pi r^2 h$. Then the weight of the kth slab is

$$(\pi r^2 \, \Delta y_k \text{ ft}^3) \cdot \left(62.4 \, \frac{\text{lb}}{\text{ft}^3} \right) = 6240\pi(\Delta y_k) \text{ lb} \qquad (r = 10 \text{ ft})$$

"Approximate" the distance of this slab from the bottom of the tank by y_k ft, where $0 \le y_k \le 20$. Therefore, the slab has to be raised $40 - y_k$ ft to the top of the tank.

The work required to raise the kth slab of water to the top of the tank is

$$W_k = \overbrace{(6240\pi \cdot \Delta y_k)}^{\text{force}} \cdot \overbrace{(40 - y_k)}^{\text{distance}}$$

The work required to pump *all* the water to the top of the tank is the sum of the work required to raise each slab:

$$W = \sum_{k=1}^{n} W_k = \sum_{k=1}^{n} 6240\pi(40 - y_k)\Delta y_k$$

which is a Riemann sum (KEY 34-6).

To find W exactly, evaluate the limit of the Riemann sum as the norm of the partition approaches 0. That limit is a definite integral (KEYS 34 and 35). Therefore

$$W = \int_0^{20} 6240\pi(40 - y) \, dy = 6240\pi \left(40y - \frac{1}{2}y^2 \right) \Bigg|_0^{20}$$

$$= 3,744,000\pi \text{ ft-lb} \approx 11,756,160 \text{ ft-lb}$$

INDEX